"A must-read if you are serious about starting, growing, or exiting a biotech. Evidence-based and data-driven. Founders and biotech executives would be remiss if this book were not on their bookshelves or part of their armamentarium."

– Joseph C. Wu, MD, PhD

Director, Stanford Cardiovascular Institute
Simon H. Stertzer, MD, Professor of Medicine and Radiology, Stanford University

"The essential guide I wish existed when I began my own journey at the intersection of science, medicine, and business. Yeung, Cruz, and Chin have masterfully distilled years of hands-on experience into a clear, actionable roadmap that demystifies the complexities of biotech valuation without oversimplifying the science or the strategy. Their frameworks—rooted in real-world case studies, rigorous analysis, and a deep understanding of both the investor and founder mindset—offer readers not just the mechanics of rNPV and SOTP, but the strategic vision needed to align innovation with capital and patient impact. Whether you are a founder navigating your first capital raise, an investor sizing up a new platform, or a scientist eager to see your discovery reach patients, this playbook will become your indispensable companion. It is a rare blend of clarity, practicality, and inspiration—a must-read for anyone serious about translating breakthrough science into real-world value."

– Jorge Cortell

Senior Advisor, Healthcare and Life Sciences, Harvard Innovation Labs

"*From Bench to Market* offers a rare and timely contribution to the biotechnology ecosystem. In an industry where technical innovation often outpaces commercial readiness, this work provides a much-needed bridge between scientific discovery and business execution. Yeung, Cruz, and Chin bring a wealth of experience to bear, presenting not only the fundamentals of valuation, deal structures, and regulatory strategy, but also the nuanced communication required to align scientists, founders, and investors around a shared vision. What sets this book apart is its clarity of purpose: to equip biotech entrepreneurs with the tools to navigate the complex journey from idea to impact. Whether you are launching your first venture or advising the next generation of innovators, this book should be regarded as essential reading."

– Ko-Chung Lin, PhD

Founder and CEO, PharmaEssentia

"A fascinating playbook for biotech maturation. Science is often the focus, yet such innovation is challenged along the pathway to integration into the complex healthcare system. Regardless of experience level, everyone must always forecast market needs, addressable markets, and strategic relationships, and, of course, maintain fidelity in planning phase 2 and phase 3 trials. Reimbursement, biosimilars, and market dynamics always pose difficulties, but identifying true clinical needs and addressing those with a pathway, or sometimes multiple pathways, increase the probability of success in delivering true solutions to improve patients' clinical needs. This is truly an important read for all stages of biotech growth, providing extremely helpful scenarios from innovation to integration."

– Peter J. Fitzgerald, MD, PhD

Co-founder and Managing Partner, Triventures
Professor Emeritus Medicine and Engineering, Stanford University

"This guide is very useful, not just for those who are deeply embedded in the biotech investment space, but for those in tangential spaces looking to break into biotech. Investing in biotech requires a different 'language' and mindset. This book does a great job providing an overview, while also equipping you with relevant terminology (e.g., 10 biotech terms you'll need to know) and things to look out for when evaluating biotech companies. Highly recommend this read!"

– Felicia Hsu, MD

Principal, BOLD Capital Partners

FROM BENCH TO MARKET

A Biotech Valuation and Forecasting Playbook

How to smartly select biotech assets that will maximize your investment potential

Tristan Yeung, MD
Jean Cruz, PhD
Yen-Po (Harvey) Chin, MD, PhD

AGMI

First Edition

Editor	Jess Lomas
Cover Designer	Ning Sen
Interior Designer	Farhan Shahid

ISBN-13: 979-8-9992926-9-8

Published in the United States

10 9 8 7 6 5 4 3 2

Disclaimer: This book is intended for informational and educational purposes only. It does not constitute financial, legal, investment, or medical advice. Readers are advised to consult appropriate professionals before making decisions based on the content herein. The author and publisher have made every effort to ensure the accuracy of the information, but do not accept responsibility for any errors or omissions, or for any loss incurred as a result of use of the material.

Every effort has been made to trace and acknowledge copyright of material used in this book. Any omissions will be rectified in future editions.

Contributions: Tristan Yeung, MD, and Yen-Po (Harvey) Chin, MD, PhD, contributed equally as co-authors. Tristan conducted the primary research and analysis, authored the majority of the content, and managed the editorial and design process. Harvey conceptualized the book project and provided strategic oversight throughout its development. Jean Cruz, PhD, contributed technical valuation expertise and provided critical industry validation to ensure accuracy and relevance.

Contents

Our Vision

As physicians, scientists, and strategists, we share a common observation: brilliant scientific innovations too often fail to reach patients—not because they lack merit, but because they lack a roadmap from bench to market. This book emerges from our collective experience at the intersection of medicine, science, and business, where we've witnessed firsthand how valuation becomes the critical bridge between breakthrough and bedside.

Each of us arrived at this realization through different paths, shaped by our distinct educational and professional journeys.

For Harvey, the realization came through his dual identity as a physician-scientist. During his MD and PhD training, he witnessed the elegance of scientific discovery and its potential to transform human health. Each research project carried the hope of a groundbreaking impact, but he gradually recognized how few innovations successfully navigated the full journey from concept to clinic to commercial scale. This observation only deepened during his time at McKinsey & Company, where he advised pharmaceutical firms, investors, and founders on complex valuation projects, experiencing firsthand the steep learning curve of translating scientific promise into business reality.

For Tristan, the insight crystallized through the heartbreaking experience of treating a young patient with a rare, aggressive form of colon cancer as a physician—a defining moment during his journey from Stanford undergraduate through Harvard Medical School to his internal medicine internship. Despite assembling a multidisciplinary team of specialists to craft a treatment plan, he found himself limited to offering only standard chemotherapy and biologics—therapies that might prolong life by months but couldn't halt the disease's progression. Watching promising therapies remain trapped in benchtop research or early-stage development while patients suffered made the translation gap painfully clear.

For Jean Cruz (JC), the insight crystallized through years of strategic work at the intersection of science and capital. Starting as a scientist before becoming a consultant at McKinsey & Company, he gained a unique perspective by helping academic spinouts navigate Series A funding, supporting global pharma players through portfolio restructuring, and valuing first-in-class assets for leading private equity firms. This positioned him between visionary founders and pragmatic investors, between scientific data and commercial narrative, repeatedly demonstrating how the right valuation framework could align all stakeholders around a common understanding of value and potential.

What unites us is the conviction that scientific progress alone cannot transform patient care. Without strategic valuation—the shared language that connects science to capital—even the most revolutionary breakthroughs risk remaining underfunded, overlooked, or lost in transition. Too often, promising innovations fail to reach patients not because they lack scientific merit, but because they lack strategic planning and a clear value articulation.

And yet, this critical topic remains almost entirely absent from traditional medical and scientific education. None of us encountered formal discussions about asset valuation, capital formation, or risk-adjusted cash flows during our training. Tristan found nothing in his medical curriculum about translating clinical insights into commercially viable treatments. Harvey's MD and PhD programs, while comprehensive in scientific methodology, offered no guidance on how to assess the market potential of research discoveries. JC's scientific training similarly lacked any framework for evaluating how laboratory findings might scale into viable therapies. The learning curve was steep when we entered the world of biotech strategy and investment—with late nights spent debating models, refining assumptions, and iterating approaches.

Through hundreds of projects across academic spinouts, venture capital firms, and global pharmaceutical companies, we began to see patterns—repeatable principles that could bring clarity to a process that too often feels opaque and subjective. Harvey identified systematic approaches to de-risking early-stage assets through his work on platform technologies and novel therapeutic modalities. Tristan observed recurring challenges when physician-inventors attempted to commercialize their ideas without strategic guidance. JC developed frameworks for valuing everything from pre-clinical assets to late-stage pipeline products, learning to balance scientific potential with market realities. Collectively, we realized that valuation, when done right, is not just about numbers but aligning vision with capital, innovation with risk, and teams with the real-world decisions they'll need to make.

This guide is our response to that gap. Together, we've distilled years of hard-earned experience into a resource that is equal parts practical and principled. You'll find frameworks, examples, and lived heuristics—tools we wish we had earlier. Whether you're a founder raising capital, an investor sizing up your next platform bet, or a translational scientist seeking to shape the arc of your innovation, this book is for you.

Our goal is to demystify biotech valuation—not by oversimplifying it, but by making it practical. We've been on both sides of the table: as scientists with a dream and as advisors entrusted with multi-billion-dollar decisions. The pages ahead contain not just valuation mechanics, but also strategic thinking, pattern recognition, and decision frameworks—the mental models that distinguish winning strategies from wishful ones.

Business acumen should never be the barrier between scientific innovation and healthcare transformation. Biotech valuation isn't merely a spreadsheet exercise—it's a language, a way to tell a story that others can believe in, invest in, and act on. It's what moves your idea out of the lab and into the hands of the people who need it most.

We hope this becomes a book you can return to as you navigate the journey from scientific discovery to patient impact. Above all, we hope this guide helps you move science forward—not just in theory, but in the world where patients are waiting.

Let's begin.

Tristan Yeung, MD

Jean Cruz, PhD

Yen-Po (Harvey) Chin, MD, PhD

01

Understanding the Biotech Business Model

Before you can accurately value a biotech product or company, you need to understand what makes these businesses unique. Unlike other industries, biotech relies on long, uncertain development cycles, heavy upfront investment, and strict regulatory oversight—factors that significantly influence how value is created, delayed, or lost.

In this chapter, we'll explore the lifecycle of a biotech company and unpack the operational realities that shape its financial profile. From early-stage research and development (R&D) to clinical trials, regulatory approval, and commercialization, we'll highlight the distinct stages biotech companies navigate—and why they matter when building a valuation model. By the end, you'll have a clear framework for how biotech companies function and the risks and milestones that impact their value over time. This foundation will support everything you learn in the chapters that follow.

Why Biotech Companies Are Different

A biotech company is a business that leverages living organisms, biological systems, and cellular and biomolecular processes to develop products and technologies that diagnose, treat, or prevent human diseases. These businesses lead innovation in areas like regenerative medicine, oncology, immunology, and neurodegeneration—advancing human health through breakthroughs in biological and medical research.

What sets biotech apart from other industries is its reliance on cutting-edge techniques and biotechnological processes that push the boundaries of biology and medicine. Unlike traditional manufacturing or service sectors, biotech companies prioritize decades-long R&D processes, which often require substantial investment and lengthy development periods before their products even have the chance to reach commercialization. This timeline encompasses rigorous preclinical testing (e.g., cell cultures and animal modeling), followed by multi-phase human clinical trials that progressively escalate in cost and complexity. Additionally, the biotech industry is highly regulated by government bodies such as the U.S. Food and Drug Administration (FDA) due to its significant clinical impact, patient safety concerns, and the associated expenditure in the healthcare system.

R&D spending is a critical factor for biotech companies—and a frequent source of financial pressure, more so than other industries. It often dominates headlines, especially when companies announce major layoffs or delays tied to clinical trial setbacks. Figure 1 illustrates global R&D intensity among the top corporate R&D spenders from 2019 to 2023 by industry sector, measured as R&D spending as a percentage of total revenue. For example, if a biotech company generates $100M in revenue and invests $19M in R&D, its R&D intensity is 19%—indicating a strategic prioritization of long-term innovation over short-term profitability. A higher R&D intensity also often indicates the increasing complexity and cost of each incremental scientific breakthrough, whether due to competitive pressures or the company operating at the frontiers of biological research where discoveries become progressively more challenging.

During this period from 2019 to 2023, the pharmaceutical and biotech sector led all industries, averaging 19% of revenue spent on R&D. Software and ICT (information and communication technologies) services followed at 14% (World Intellectual Property Organization 2024). Notably, pharma and biotech experienced a significant dip in 2022 from 2020, with R&D investment falling by 2%, but rebounded by 3% in 2023 as pandemic-related disruptions from COVID-19 eased.

R&D INTENSITY BY SECTOR, 2019 TO 2023

- — Biotech and Pharmaceuticals
- Software and ICT Services
- ICT Hardware and Electrical Equipment
- Automobiles
- Travel, Leisure and Personal Goods
- Healthcare Equipment and Services
- Construction and Industrial Metals

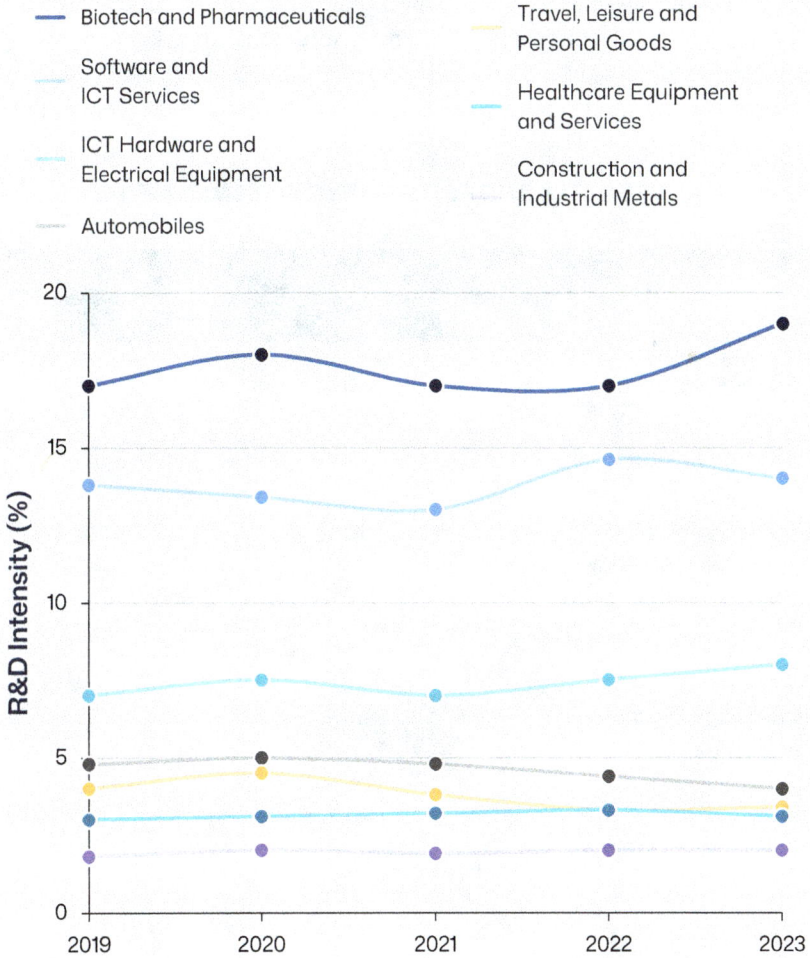

Note: R&D intensity is defined as the amount of global R&D spending as a percentage of total revenue of the top corporate R&D spenders.

Source: World Intellectual Property Organization 2024

Figure 1.   Global R&D Intensity by industry sector from 2019 to 2023.

To manage risk and avoid financial failure, biotech companies typically build a diverse portfolio of product candidates—each at different stages of preclinical or clinical development. For example, a biotech company might have one drug in early preclinical testing, another entering clinical trials, and a third already approved for use. Managing a mix like this spreads risk and ensures that setbacks in one area don't threaten the entire business. These risk-return dynamics are illustrated in Table 1, which presents the approximate risk and returns of different investment options.

FAILURE RATES AND RETURNS BY INVESTMENT TYPE

Investment Type	Failure Rate	Return on Investment
U.S. Treasury Bonds	~0%	4%
Corporate Bonds	5-30%	8%
Established Public Companies (S&P 500)	10%	10%
Tech Companies (NASDAQ)	50%	20%
Early-Stage Biotech Companies	80%	50%

Note: Provided figures are estimates for illustrative purposes only and should not be relied upon for investment decisions.

Table 1.   Impact of investment type on failure rate and expected return on investment.

Advancing these programs, running clinical trials, and sustaining operations all require substantial capital. Every dollar counts, and—as the saying goes—time is money. Delays can be costly, both financially and strategically, which is why maintaining adequate cash flow and attracting investors to fund R&D and company operations are both critical elements for survival and success in building a biotech business

These challenges—substantial upfront capital requirements, extended development timelines often spanning decades, and the intense investor scrutiny of R&D milestones—make it crucial for biotech companies to demonstrate justify and demonstrate the evolving value of their scientific assets. Some unique challenges include:

- **Limited Transparency:** Many biotech companies remain private for extended periods, making comparable financial data scarce and difficult to access.

- **Technical Complexity:** The intricate nature of scientific IP and technology platforms often requires specialized knowledge to properly assess value.

- **High Uncertainty:** Regulatory hurdles and clinical development factors introduce significant unpredictability into the valuation process.

- **Risk-Reward Profile:** The sector offers exceptionally high potential returns balanced against substantial investment risks.

- **Extended timelines:** The lengthy investment cycle from drug discovery to commercialization requires sophisticated long-term forecasting models.

That's where forecasting and valuation come in. The process helps begin the process of understanding the business model, market, and competitive landscape, which is particularly useful for early-stage companies with no established product or track record.

Valuation as a Strategic Tool

Bringing a new biotech product to market is a long, uncertain journey—one often constrained by high development costs and significant risk. Forecasting and valuation play a crucial role in navigating this landscape. By quantifying the potential of developing assets and demonstrating why such significant investments are necessary, valuation models help justify large investments, guide decision-making, and strengthen a company's strategic positioning.

Whether a company is assessing its own pipeline or negotiating with investors, these valuations must be backed by solid reasoning. This is where forecasting techniques and valuation models come into play.

Biotech valuation models estimate the potential value of early- and commercial-stage products, serving three key purposes:

1. **Quantifying Potential Value:** Valuation models help estimate the financial potential of biotech products. This is crucial for attracting investment and securing funding for further development.

2. **Informed Decision-Making:** These models provide insights that support high-impact decisions, including whether to advance clinical development, launch new product evaluation, pursue licensing deals, or initiate mergers and acquisitions.

3. **Risk Assessment:** Valuation tools factor in the probability of success and account for development risks, which are essential in an industry with high failure rates. This helps investors and teams understand the relationship between risk, cost, and return.

Understanding how valuation works is only part of the equation. At its core, valuation hinges on assumptions and quantifying risk, ultimately leading to informed assessments about assets or companies. To be truly effective, these models must meet the needs of the people making the decisions—whether they're in the lab, the boardroom, or the investor's chair.

Let's briefly explore the key players who depend on biotech valuation models and what they're trying to accomplish.

Who Uses a Biotech Valuation Model?

A strong, data-driven valuation model is one of your most powerful tools when working with stakeholders across the biotech ecosystem. It serves as the crucial link between scientific discovery and business success, which translates complex research achievements into financial terms that promote communication between scientists, executives, and investors.

Key users include:

- **Management Teams, Board Members, R&D Leaders, and Company Directors:** Valuation models act as strategic roadmaps, helping leadership allocate limited resources to the most promising projects to maximize success.

- **Acquisition and Licensing Teams:** Business development professionals and investment bankers use valuation models to assess potential deals and determine whether acquiring or licensing a new technology is a sound financial decision.

- **Biotech Investors and Shareholders:** Investors rely on valuation models to predict a company's or project's potential performance. These models provide a clearer understanding of expected returns and risks, helping investors decide whether to commit capital or explore other opportunities.

By integrating valuation into strategic planning, biotech companies and investors alike can navigate the complexities of drug development and investment with confidence and clarity.

A Peek into Biotech Investment Trends

Headlines about biotech megadeals may catch your eye—but the real value lies in understanding what drives these transactions beyond their impressive figures. When Biotech Company A secures a multimillion-dollar partnership with Biotech Company B, that investment reflects a calculated assessment of scientific potential, market need, and competitive positioning rather than simply deep pockets or industry hype.

Investment patterns across therapeutic areas reveal which scientific approaches and disease targets are gaining credibility with investors conducting rigorous due diligence. By recognizing which modalities or disease areas are attracting capital—and more importantly, why—you gain critical intelligence that can help you position your own assets more strategically, articulate your value proposition more convincingly, or anticipate market movement before it becomes obvious to less analytical observers.

Figure 2 illustrates the leading therapeutic areas by Seed or Series A funding (early-stage capital) and total deal value in 2023 (JPMorgan Chase & Co. 2023). When evaluating these numbers, it's important to remember that valuations are typically built on assumptions supported by industry benchmarks and well-thought-out reasoning.

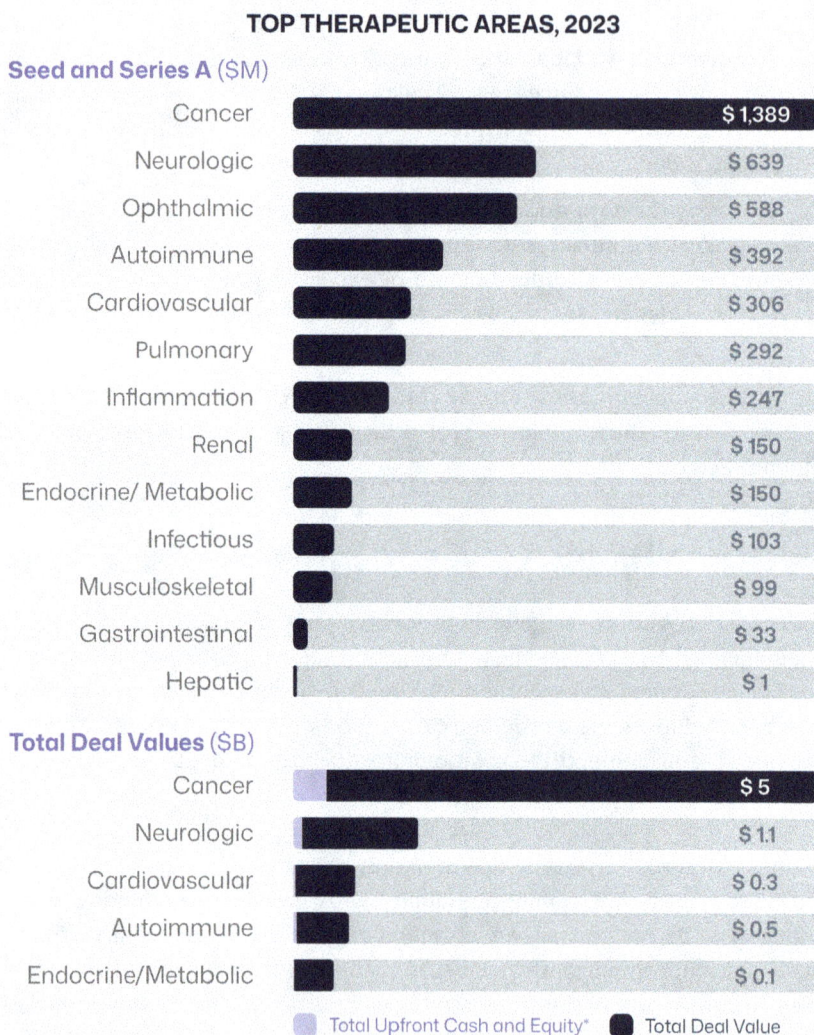

TOP THERAPEUTIC AREAS, 2023

Seed and Series A ($M)

Cancer	$ 1,389
Neurologic	$ 639
Ophthalmic	$ 588
Autoimmune	$ 392
Cardiovascular	$ 306
Pulmonary	$ 292
Inflammation	$ 247
Renal	$ 150
Endocrine/ Metabolic	$ 150
Infectious	$ 103
Musculoskeletal	$ 99
Gastrointestinal	$ 33
Hepatic	$ 1

Total Deal Values ($B)

Therapeutic Area	Value
Cancer	$ 5
Neurologic	$ 1.1
Cardiovascular	$ 0.3
Autoimmune	$ 0.5
Endocrine/Metabolic	$ 0.1

Total Upfront Cash and Equity*   Total Deal Value

Therapeutic areas receive total upfront cash and equity investments, but some areas may be too small to display on this chart.

Source: JPMorgan Chase & Co. 2023

Figure 2. Top therapeutic areas by Seed and Series A funding and total deal value in 2023.

Clear patterns are emerging in early-stage biotech investments, particularly in Seed and Series A funding rounds. In 2023, cancer led globally, attracting $1.39B in early-stage funding. It was followed by neurologic disorders ($639M), ophthalmic diseases ($588M), and autoimmune conditions ($392M). These figures reflect strong investor interest in areas with significant unmet clinical needs and complex treatment challenges, such as oncology, neurology, ophthalmology, and rheumatology.

But early-stage funding is only one part of the picture. As therapies progress and companies strike licensing deals with larger biotech or pharma partners, deal values can skyrocket. Cancer again leads, with a total deal value of $86.1B. Neurologic disorders follow with $21.1B, while cardiovascular and autoimmune diseases have values of $10.3B and $9.5B, respectively.

Furthermore, we can break down the market by therapeutic modality, revealing what specific biotechnologies investors prioritized. According to the latest industry analysis, global investments in biotech Seed and Series A funding totaled close to $9.5B in 2024 (JPMorgan Chase & Co. 2025), with major investments focused on the following areas:

1. **Biologics (Antibody, DNA, RNA, Protein, etc.) ($4.1B):** Biologics continue to dominate investment flows as these therapeutic modalities offer proven methods for treating complex diseases. This category encompasses antibody therapies, nucleic acid-based treatments, and protein therapeutics that leverage the body's natural biological processes.

2. **Small Molecule ($2.5B):** Despite the rise of biologics, small molecule drugs remain a cornerstone of drug development, offering advantages in oral bioavailability, manufacturing scalability, and broad therapeutic applications across multiple disease areas.

3. **Genomics, Sequencing, Screening, etc. ($1.8B):** Investment in genomic technologies reflects the growing importance of precision medicine and personalized therapeutics. These platforms enable better patient stratification, biomarker discovery, and targeted treatment approaches.

4. **Cell Therapy ($470M):** While representing a smaller absolute investment compared to other modalities, cell therapy continues to attract significant interest as investors support advances in engineered therapeutic cells and their delivery mechanisms for treating previously intractable conditions.

5. **Gene Therapy and Vectors ($296M):** Gene therapy investments support development of improved delivery systems and therapeutic approaches for correcting genetic disorders, though funding levels reflect investor caution related to challenges in translating these promising technologies to commercial success.

The 2024 data reveals that investors maintained confidence in established therapeutic modalities while continuing to support emerging technologies, with biologics and small molecules capturing the majority of early-stage investment activity. However, gene and cell therapy saw some pullback in deal flow as investors and partners turned to more standard modalities (JPMorgan Chase & Co. 2025).

So, why does this matter? Understanding where capital is flowing helps you identify the technologies and therapeutic areas with the greatest perceived value. For founders, researchers, and investors, understanding these trends can guide everything from choosing your lead indication to building realistic financial models based on what the market is already funding. Used well, this information can help you shape your assumptions, refine your pitch, and make more strategic decisions about where to focus your efforts. Convincing key stakeholders means securing more funding, forging stronger partnerships, and ultimately increasing your chances of bringing groundbreaking therapies to patients.

After examining where investment capital is flowing in biotech and its importance to building a reliable valuation model, it's essential to understand the strategic relationships that underpin successful companies in this sector. These key partnerships not only influence investment decisions but often determine a biotech firm's ability to translate promising science into commercial success.

Understanding the Biotech Ecosystem

Biotech companies don't succeed in isolation. Their progress—and long-term success—depends on building and maintaining strong strategic relationships. From securing funding and partnering with manufacturers and clinical developers to collaborating on cutting-edge research, these connections shape a product's clinical positioning, commercial strategy, and pathway to approval. Nurturing these partnerships or alliances is essential for navigating the biotech industry and achieving long-term success.

Figure 3 offers a simplified overview of the major players in the biotech and broader drug development ecosystem. It illustrates how key stakeholders—patients, doctors, payers, commercial pharma, emerging biotech firms, and various investor groups—interact throughout the product lifecycle. The flowchart also highlights the ongoing cycle of investment and returns in the pharmaceutical industry, focusing on the exchange of cash, technology, and expertise among these groups.

BIOTECH INDUSTRY ECOSYSTEM AND VALUE CHAIN

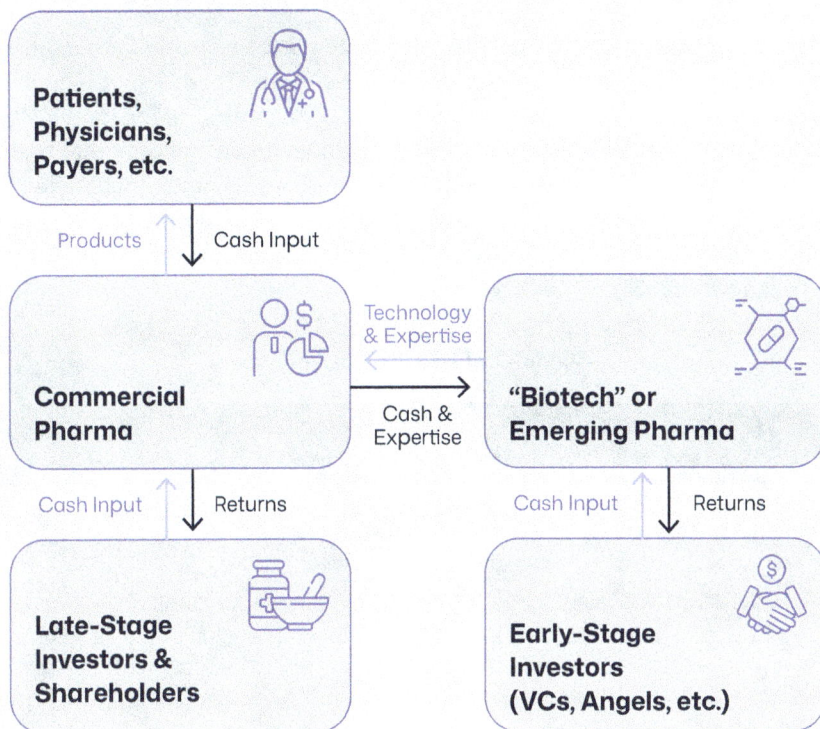

Figure 3. *Overview of the biotech industry ecosystem and value chain.*

From the perspective of early biotech companies, there are usually two main types of partners: early-stage investors (VCs, angel investors, etc.) and big pharma companies. Early-stage investors are crucial because they provide the funding needed to kickstart new projects. Investors in biotech focus on identifying market needs, rigorously evaluating new technologies or assets, and assessing their potential to drive impact and generate strong cash flow. On the other hand, partnerships with big pharmaceutical companies offer more than just financial support. These companies bring valuable research expertise, resources, and access to their development pipelines. In return, they get rights to the biotech company's cutting-edge technologies and the specialized knowledge behind them.

Furthermore, commercial pharma companies rely on late-stage investors as well as their shareholders to sustain and expand their operations. These investors provide capital for further research, acquisitions, and global market expansion. They also help ensure that the pharmaceutical company remains competitive and continues to innovate.

How do biotech companies secure funding?

While venture capital (VC) firms often dominate the conversation around early-stage funding, many other investment sources exist. These can generally be categorized as equity-based or non-equity-based. Regardless of the funding type, a solid valuation model is essential as it acts as a common language that bridges companies with the capital they need. Table 2 outlines common funding options available to any early-stage business (Chitale et al. 2022). For more information regarding funding sources, please refer to University Lab Partners (2021), Murphey (2023), and Brex (2025).

COMMON FUNDING SOURCES

Sources of Capital	
Non-equity-based	**Equity-based**
⋏ Federal Small Business Innovation Research (SBIR) grant programs	⋏ Friends and family investment
⋏ Federal Small Business Technology Transfer (STTR) grant programs	⋏ High-net-worth or angel investors
⋏ Entrepreneurship competition awards	⋏ Family offices
⋏ Crowdfunding	⋏ Institution internal investment funds
⋏ Foundation grants	⋏ Foundation grants
⋏ Collaborations with larger or more established companies	⋏ Venture capitalists and other "institutional" investors

Source: Chitale et al. 2022

Table 2.   Common funding sources for biotech companies.

Key Concepts of a Biotech Company

Before diving into the specific aspects that are important in forecasting and valuation, let's first explore the broader context of a biotech company's journey. We'll first clarify some key biotech concepts and then outline the essential stages of a biotech company's lifecycle, while focusing on the crucial factors that influence its growth and success.

Defining Key Concepts

To understand a biotech company's strategy and market opportunity, it helps to start with a few foundational terms. These core concepts shape how companies define their focus, attract investment, and communicate value.

Therapeutic Area

A Therapeutic Area (TA) is a broad field of medicine that focuses on treating a related group of diseases or medical conditions. It defines the medical category a product or pipeline belongs to and is central to how companies position themselves in the market. For example:

Oncology

Addresses malignancies across all tumor types, including solid tumors (lung, breast, colorectal) and blood cancers (leukemia, lymphoma).

Neurology

Focuses on diseases of the brain, spinal cord, and nervous system, including Alzheimer's disease, Parkinson's disease, epilepsy, multiple sclerosis, and stroke.

Cardiology

Concentrates on diseases of the heart and circulatory system, including coronary artery disease, heart failure, irregular heart rhythms, and high blood pressure.

Mechanism of Action

A Mechanism of Action (MoA) describes the specific biochemical pathway or molecular process by which a drug produces its therapeutic effect. Understanding a drug's MoA is critical for evaluating its potential efficacy, safety profile, and suitability for specific patient populations. Common MoAs include:

- **Hormone Mimicry:** GLP-1 receptor agonists, such as Ozempic and Wegovy, mimic the effects of GLP-1, a hormone that helps regulate blood sugar levels. By increasing insulin production and decreasing glucagon secretion, these drugs help lower blood sugar levels.

- **Protein Inhibition:** Some MoAs block or inhibit proteins that play a pivotal role in an illness. For example, PD-1 inhibitors prevent PD-1 proteins from suppressing T-cell activity, enhancing the immune system's ability to recognize and attack cancer cells.

- **Enzyme Targeting:** Certain drugs work by inhibiting specific enzymes that drive disease progression. For instance, tyrosine kinase inhibitors (TKIs) block enzymes involved in cancer cell signaling pathways, thereby preventing tumor growth.

Therapeutic Modality

A Therapeutic Modality (TM) refers to the specific technology platform, delivery mechanism, or treatment approach used to administer a therapeutic intervention, encompassing the fundamental method by which a treatment exerts its biological effect. Different technologies offer unique advantages, like improved efficacy, fewer side effects, or better precision in targeting diseases. Key examples include:

Small Molecule Drugs

Chemically synthesized compounds with relatively simple molecular structures that can penetrate cells and interact with specific intracellular or extracellular biological targets. These oral or injectable medications typically have predictable pharmacokinetics (how the body absorbs, distributes, breaks down, and eliminates the drug) and are manufactured through traditional chemical synthesis processes.

Traditional Biologics

Large, complex therapeutic molecules derived from living organisms or produced using biological manufacturing systems. Examples include recombinant proteins, monoclonal antibodies, vaccines, blood products, and other biomolecules that primarily target extracellular proteins and cell surface receptors. Because of their substantial molecular size, structural complexity, and vulnerability to degradation by digestive enzymes, traditional biologics generally require administration through injection, infusion, or other routes rather than oral delivery.

Cell and Gene Therapies

Advanced treatment approaches that directly modify or replace cellular and genetic components to treat disease. Cell therapy involves introducing living cells (either the patient's own modified cells or donor cells) into the patient, while gene therapy delivers genetic material to correct defective genes or provide new cellular functions.

RNA-Based Therapies

Therapeutic approaches that use various forms of RNA to treat disease by targeting specific genetic sequences or cellular processes. These include messenger RNA (mRNA) therapies, small interfering RNA (siRNA), antisense oligonucleotides, and microRNA therapeutics that can modulate protein production, silence disease-causing genes, or provide instructions for therapeutic protein synthesis.

> **Note:** Cell and gene therapies and RNA-based therapies are categorized as biologics by regulatory agencies such as the FDA due to their biological origin, complex manufacturing processes involving living systems, and large molecular structures that distinguish them from traditional small molecule drugs.

Therapeutic Indication

A Therapeutic Indication (TI) refers to the specific use of a drug to treat a particular condition. For a drug to be marketed for an indication, it must receive approval from the FDA, ensuring it's proven safe and effective for that specific condition. Indications can be thought of as subcategories within a therapeutic area. For instance:

Oncology

Breast cancer, lung cancer, melanoma.

Neurology

Alzheimer's disease, Parkinson's disease, ALS.

Musculoskeletal

Rheumatoid arthritis, osteoarthritis.

In summary, these four core concepts work in tandem to shape a biotech company's strategy, competitive positioning, and value proposition:

01 **Therapeutic area** defines the broad medical field the company is targeting, helping determine market size and competitive dynamics.

02 **Mechanism of action** explains how the drug exerts its effects, influencing decisions on combination therapies, potential side effects, and market differentiation.

03 **Therapeutic modality** shapes the drug development strategy, competitive differentiation, and scalability potential while determining regulatory complexity and manufacturing requirements.

04 **Therapeutic indication** specifies the precise conditions the product aims to treat, sharpening the company's focus to reach specific patients or providers, guiding resource allocation for partnerships, and shaping financial projections.

U.S. Drug Approval Pathways

Before exploring the lifecycle of a biotech company, it's important to understand how new drugs gain regulatory approval. Agencies such as the FDA evaluate whether investigational treatments are safer or more effective than existing options before granting approval.

Before a drug can enter human clinical trials, the sponsor—typically a biotech or pharmaceutical company—must submit an Investigational New Drug (IND) application to the FDA (U.S. Food and Drug Administration 2014). This application includes preclinical data (e.g., laboratory and animal studies), the proposed clinical trial design, and details on manufacturing processes to ensure the drug can be safely tested in humans.

Once human clinical trials are complete, the sponsor submits a New Drug Application (NDA) for small-molecule drugs or a Biologics License Application (BLA) for biologic therapies, which undergo a full FDA review (U.S. Food and Drug Administration 2014).

To expedite the approval process for drugs that address serious conditions with high unmet medical needs, the FDA has established accelerated pathways, which can impact the return on investment by reducing costs and increasing the period of exclusive commercialization. The FDA offers several key designations (U.S. Food and Drug Administration 2023):

- **Fast Track:** Designed to expedite the development and review of drugs targeting serious conditions with high unmet medical needs. Sponsors benefit from more frequent interactions with the FDA and can submit regulatory filings on a rolling basis.

- **Breakthrough Therapy:** Granted to drugs that may offer substantial improvements over existing treatments. Like Fast Track designation, it allows for increased FDA interactions and rolling submissions but also provides additional support to accelerate development.

- **Accelerated Approval:** Allows drugs for serious conditions to be approved based on surrogate endpoints—measurable markers that are reasonably likely to predict clinical benefit—rather than waiting for long-term outcomes. This pathway is common in diseases like cancer, where tumor shrinkage can serve as an early indicator of effectiveness.

- May allow for conditional approval after a Phase 2 clinical trial, with a Phase 3 trial required to confirm efficacy and safety. If confirmatory trials fail, the FDA can modify or withdraw approval.

- **Priority Review:** Shortens the FDA's review period to six months (compared to the standard 10 months) for drugs that may offer a significant therapeutic advantage over existing treatments.

While not a formal designation, sponsors may design combined or hybrid clinical trials (e.g., Phase 1/2 or Phase 2/3) to reduce the time and resources for drug development, subject to FDA approval.

These approaches can streamline patient recruitment, enhance data collection efficiency, and shorten overall development timelines while maintaining compliance with regulatory requirements. Agencies like the FDA rigorously evaluate clinical trial designs and outcomes to ensure they meet the necessary standards for approval.

10 Biotech Terms You'll Need to Know

Before reading the next chapters, familiarize yourself with these essential terms that form the core concepts of biotech valuation. These will appear throughout this guide as we explore how to assess a biotech product's financial potential.

Cost of Goods Sold (COGS)

01

The direct costs attributable to producing a drug, including raw materials and manufacturing expenses, which affects profit margins.

Discounted Cash Flow (DCF)

02

A valuation method that estimates the Net Present Value (NPV) of an investment by projecting future cash flows and adjusting for the time value of money and associated risks.

Early-Stage Biotech

03

Companies focused primarily on research and development with no or limited commercial products, characterized by high risk and significant upside potential.

04 Probability of Success

The likelihood that a drug candidate will successfully complete a development phase and transition to the next, typically estimated based on therapeutic area.

05 Loss of Exclusivity (LOE)

The point when a drug's patent protection or market exclusivity expires, allowing competitors to introduce generic or biosimilar alternatives.

06 Market Exclusivity

The period during which a company has exclusive rights to market an approved drug, directly impacting revenue potential before generic competition.

07 Research and Development (R&D)

The phase focused on discovering and developing new products, representing significant investment in early-stage biotech companies.

08 Selling, General, and Administrative (SG&A)

Expenses related to marketing, distribution, and business operations that significantly impact a biotech product's profitability.

09 Serviceable Addressable Market (SAM)

The portion of the total market that a biotech product can realistically target given its specific therapeutic profile, pricing strategy, and competitive positioning.

10 Total Addressable Market (TAM)

The total revenue opportunity for a product if it could capture 100% of its relevant market, representing the upper boundary of market potential.

Lifecycle of a Biotech Product

The journey from preclinical or clinical development to market launch and commercialization is complex, with each stage presenting its own challenges and milestones. Biotech companies often develop multiple products simultaneously, all of which must progress through these key stages. For example, one product may be in the clinical development stage while another may be in the growth stage of commercialization. Figure 4 outlines the key phases in the development of a single biotech product. We'll explore each of these briefly in the section that follows. As previously discussed, a biotech company can have many products at different points in the lifecycle, ranging from early research to full-scale commercialization, which is not depicted in the figure for simplification purposes.

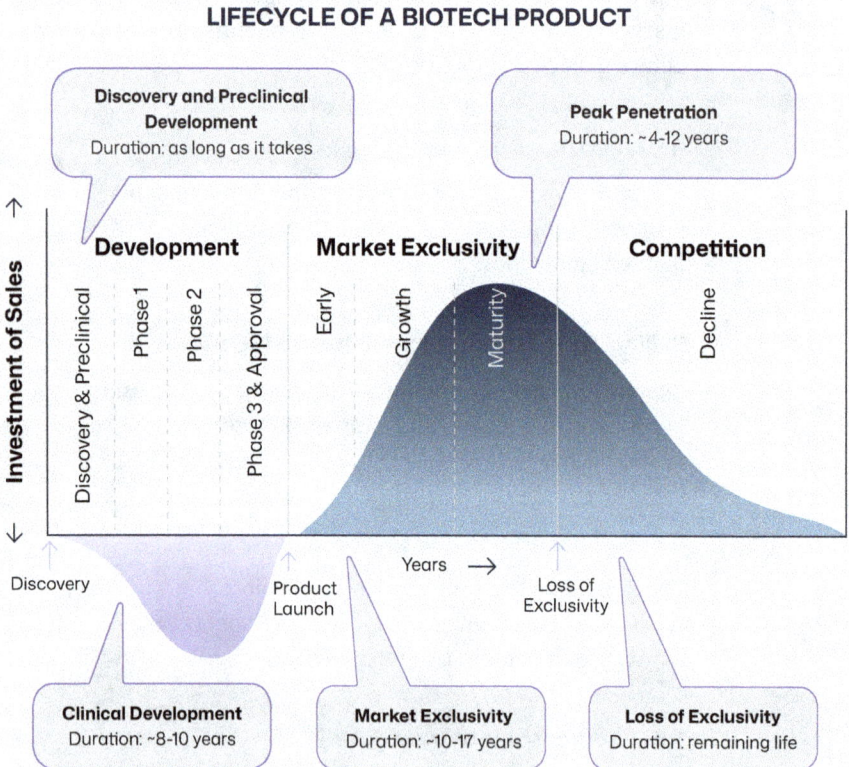

Figure 4. *Biotech product lifecycle from preclinical to clinical development, market exclusivity, and entry of major competitors.*

Discovery and Preclinical Development

Duration: As long as needed (average 5.5 years)

In the early stages of drug development, the focus is on discovery and preclinical research to identify a promising biotech product or drug candidate. To ensure the new therapeutic is safe for human testing, researchers conduct toxicology tests and other preclinical studies. These studies form the basis of the IND application submitted to the FDA to request initiation of clinical trials.

There is no set timeline for this phase because it depends on how long it takes to identify and optimize a viable candidate. On average, the process includes:

- ~1 year to identify a drug target
- ~1.5 years to find a lead candidate
- ~2 years for optimization
- ~1 year for animal testing

This foundational phase sets the stage for everything that follows. It is where scientific feasibility is established, initial risk is assessed, and early data begins to shape regulatory strategy and long-term development plans.

(Schlander et al. 2021; Paul et al. 2010)

Clinical Development

Duration: ~8–10 years (average of 9 years)

Once a company identifies a promising drug candidate and receives IND approval, the product enters clinical development. This stage involves testing the drug in human subjects to evaluate its safety, efficacy, and performance relative to existing treatment standards. Clinical development typically takes an average of 9 years, depending on the complexity of the trials and the challenges encountered.

In the U.S., this process is divided into three main phases:

- **Phase 1 (~1.5 years):** This is the safety check. A small group of healthy volunteers or patients try the drug so researchers can make sure it's safe to use.

- **Phase 2 (~3.6 years):** Next, the focus shifts to how well the drug works. A bigger group of patients takes the drug, and safety is still a top priority.

- **Phase 3 (~3.3 years):** This is the pivotal phase. Thousands of patients participate in comparing the drug against existing treatments or gold-standards. Effectiveness, side effects, and long-term outcomes are closely monitored.

Each phase builds on the last, producing evidence across a range of endpoints to support regulatory approval. Once Phase 3 trials are successfully completed, the company can submit an NDA for small-molecule drugs or BLA for biologic therapies to the FDA for final review. If approved, the drug can be commercialized and launched into the market for widespread patient use.

(DiMasi et al. 2016; Jayasundara et al. 2019; QLS Advisors 2021; Brown et al. 2022; Sertkaya et al. 2024)

Market Exclusivity

Duration: ~10–17 years (average 14 years)

Once the drug candidate successfully passes the requisite clinical trial and receives FDA approval to be commercialized in the U.S., the company can start selling the new therapeutic as the sole manufacturer (i.e. market exclusivity) for an average of 14 years, depending on initial patent submission, clinical trial lengths, specific drug application, and federal regulations. During this period, the company is protected from competition, allowing it to recoup R&D costs and potentially turn a profit. Exclusivity is crucial for driving revenue, which investors seek to maximize their returns as it provides a period of market dominance before competitors can introduce alternatives.

(Grabowski et al. 2021; Wang et al. 2015; Rome et al. 2021)

Peak Penetration

Duration: ~4–12 years (average 8 years)

Following launch, a biotech product typically enters a growth phase where sales steadily increase—culminating in peak market penetration. This stage represents the product's commercial high point, often reached around 8 years after launch. To maximize sales and reach this peak, the company will employ strategic marketing and sales strategies to maximize revenue in the setting of exclusive market access. This involves building a robust sales team, executing targeted marketing campaigns, and collaborating closely with healthcare providers and patients to drive product adoption.

Achieving peak penetration requires a coordinated effort across commercial, medical, and marketing functions:

- **Sales and Marketing Strategy:** Building a strong sales force and executing targeted campaigns to educate physicians, engage payers, increase patient awareness, and support prescribing behavior.

- **Medical Affairs & Medical Science Liaisons (MSLs):** MSLs serve as the bridge between the company and key opinion leaders (KOLs), sharing clinical data, answering product questions, and supporting safe and informed product use.

Together, these teams drive not only commercial success but also ensure the product's credibility and long-term value in the market.

(Robey and David 2017; Teramae et al. 2020)

Loss of Exclusivity (LOE)

Duration: Remaining life of the therapy

Eventually, every biotech product faces the end of its market—typically when its patents expire or exclusivity protections lapse. At this point, generics or biosimilar competitors may enter the market after approval from abbreviated regulatory pathways, often at discounted and competitive prices. Though "generics" is sometimes used as a blanket term for all post-exclusivity competitors, this guide distinguishes between two important categories:

- **Generics:** These are pharmaceutical equivalents of small-molecule drugs with well-defined chemical structures. True generics must be bioequivalent to the reference product, containing identical active ingredients in the same dosage form, strength, and route of administration. They must demonstrate the same pharmacokinetic profile, ensuring they provide identical therapeutic effects while meeting the same quality and safety standards.

- **Biosimilars:** These are highly similar versions of biologic medicines—usually complex products derived from living cells or organisms rather than chemical synthesis. Examples include monoclonal antibodies, recombinant proteins, and advanced cell therapies. Unlike generics, biosimilars cannot be exact copies due to the inherent variability of biological manufacturing processes. Instead, they must demonstrate comparable efficacy, safety, and immunogenicity profiles through comprehensive analytical studies and clinical trials, with any minor structural differences having no clinically meaningful impact on patient outcomes.

This is the product's "twilight" phase, when competition increases, prices drop, and sales begin to decline. While patents may protect the product during development and early commercialization, market exclusivity—enforced through a government-approved drug application—prevents competitors from launching similar products for a defined period after approval. Although patents and exclusivity overlap, the expiration of either typically marks the end of peak revenue and the beginning of the loss-of-exclusivity period.

To counter this inevitable decline, companies often pursue label expansion, aiming to extend the product's market potential by securing approval for new indications, formulations, or patient populations. This strategy can help breathe new life into the product, capture additional market share, and maintain revenue streams as competition increases. At the same time, leading biotech firms invest in pipeline innovation, ensuring new therapies are ready to offset the financial impact of LOE.

(U.S. Food and Drug Administration 2020; VCL Solutions 2020)

Approaches to Biotech Valuation

Now that we've explored the lifecycle of a biotech product—from early development through to LOE—it's time to look at how that journey translates into value. Whether you're assessing a single drug or an entire company, understanding how to approach valuation is key to making smart, informed decisions.

At its core, biotech valuation aims to answer three deceptively simple yet profound questions that drive investment and development strategy:

- What is the current market value of the product or company?
- How will its value—and associated risks—evolve throughout each phase of development?
- Is the product or company positioned for long-term financial viability in a competitive landscape?

While these questions appear simple, the process of valuation is a nuanced and calculated process. The foundation of many biotech valuation approaches lies in Discounted Cash Flow (DCF) analysis—a method that estimates an asset's intrinsic value by projecting its future cash flows and discounting them back to present value.

However, how you apply DCF principles will depend on whether you're evaluating a single biotech product or a company with a portfolio of products. This distinction matters because your valuation purpose—whether licensing a single asset or evaluating an entire company for mergers and acquisitions (M&A)—determines which method to use:

Valuing a Single Biotech Product
(Licensing, Private Equity, Hedge Funds)

Net Present Value (NPV) serves as the primary output of DCF analysis—the sum of all discounted cash flows (both inflows and outflows) over the investment period. NPV is widely used across industries to evaluate investments by determining whether the present value of all expected cash flows results in a positive or negative value. While NPV is reliable for commercial companies with predictable cash flows, it is less effective for the high-risk, early-stage products found in the biotech industry. Therefore, a more specialized approach is necessary.

Risk-adjusted Net Present Value (rNPV) adapts the traditional NPV model by incorporating the uncertainties associated with the clinical development and commercialization process. Milestones such as clinical trial outcomes and FDA approvals can significantly impact the value of early-stage biotech products, which can determine financial viability. By incorporating these risks, rNPV delivers a more reliable and data-driven valuation that reflects the inherent risk of R&D in the biotech industry.

Valuing a Biotech Company

(M&A, Venture Capital)

The Sum-of-the-Parts (SOTP) method is a valuable approach for assessing biotech companies with a portfolio of products. Building on the rNPV method, SOTP evaluates each product individually, factoring in market opportunity, revenue potential, and associated risks, before aggregating these valuations along with company-wide financial variables to estimate the company's total value. This approach is particularly useful as it enables a detailed analysis of multiple assets, helping to optimize resource allocation and strategic decision-making within the broader context of the company's portfolio.

What to Keep in Mind

We've just explored the fascinating lifecycle of biotech products, from discovery to market launch and beyond. Each stage brings its own set of challenges and timelines, and understanding these moving parts is crucial when you're trying to put a dollar value on potential.

The valuation approach we'll cover in this guide will help make sense of this complex journey, but it's worth noting how the industry typically handles this challenge. Many investment banks maintain dedicated in-house analysis teams that value products using analog case studies and expensive proprietary databases containing detailed market analysis, product dynamics, and industry benchmarks. However, our guide aims to make valuation accessible while being data-driven and rigorous by utilizing readily available resources and benchmarks that help you build robust risk-adjusted models without requiring very costly specialized resources.

As we continue our journey together, we'll dive deeper into practical methods for valuing both individual assets and entire company portfolios using rNPV and SOTP approaches. Think of the coming chapters as your toolkit—filled with industry benchmarks, calculation methods, and real-world examples that will help you build valuation models that actually make sense in the wild and unpredictable world of biotech.

Here's a sneak peek of fundamental topics we'll cover:

Chapter 2: Market Opportunity

Learn how to assess demand, map the competitive landscape, and understand pricing power.

Chapter 3: Product Dynamics

Explore market entry strategies, the impact of exclusivity loss, and the cost of getting a product to market.

Chapter 4: Research and Development

Examine the true timelines, costs, and probabilities behind clinical development success.

Chapter 5: Operating Expenses, Discount Rates, and Payment Structures

Unpack the operational costs, deal terms, and financial factors that shape a product's value.

All of these components will come together in **Chapter 6: Biotech Valuation in Action**, where we'll walk through a full biotech valuation case study. To make this as practical as possible, we've included a pre-built, user-friendly Excel model designed to simplify the learning process. With intuitive inputs and automated calculations, you can focus on insights rather than complex formulas—just enter your numbers and let the model do the heavy lifting.

By the end of this book, we hope you will not only have a solid grasp of your biotech project's financial potential, but you'll also gain the confidence to communicate its value to investors and stakeholders. Whether you're advancing cutting-edge research or preparing for funding discussions, this knowledge will be a game-changer.

BEFORE YOU WRAP UP

Take a moment to reflect on what you've learned. These key takeaways and questions are designed to help you connect the concepts in this chapter to your own goals and identify areas you'd like to explore further.

Key Takeaways

- Biotech companies are unique due to their reliance on cutting-edge science, long research and development (R&D) timelines, and high regulatory oversight.

- Understanding biotech's business model—especially its high cost, risk, and need for early funding—is essential for anyone seeking to forecast or evaluate its potential.

- Valuation isn't just for investors; it's a critical tool for founders, researchers, and company leaders who need to make data-informed decisions and attract funding.

- Risk-adjusted Net Present Value (rNPV) and Sum-of-the-Parts (SOTP) are two key valuation methods you'll learn to apply throughout this guide.

- Identifying a product's stage in the lifecycle—from discovery to loss of exclusivity (LOE)—helps shape expectations around value, cost, and opportunity.

Reflection Questions

1. What surprised you most about the biotech company lifecycle?

2. Based on what you've read, how would you explain the value of a biotech product to someone outside the industry?

3. Which stage of the lifecycle do you think is most critical for valuation— and why?

4. Are there companies or innovations you've come across that you now see differently after learning about rNPV or SOTP?

5. What would you like to better understand as you move into the next chapter?

02
Market
Opportunity

When launching a new treatment, it's not just about sizing up the potential market but gauging the current market and demand. The real value lies in understanding how your innovation can excite, engage, and drive meaningful impact in the market. Success comes from connecting that early excitement to a clear plan that creates value from your developing biotech product, which will become the foundation to creating a sustainable business model. Key considerations include:

1. Does your innovation address an unmet need?

2. What competing products already exist?

3. How will your product be priced?

By evaluating both market size and industry trends, you gain a clearer understanding of how your product could impact patient care, differentiate itself from competitors, and drive future revenue growth. This insight informs critical decisions on investment, product development, and commercialization strategy, helping shape its success.

Fundamentals of Market Assessment

· · · · · · · · ·

Accurately estimating the market size is essential for understanding both the product's potential impact and its revenue prospects—key factors that are critical for attracting investors and obtaining funding. There are two main ways to figure out how many people might use a new product: we can either count potential patients who need treatment (called the "bottom-up" approach) or look at similar products already being sold (the "top-down" approach). In this guide, we'll use the first method—counting potential patients—because most new products don't have similar alternatives already available for comparison.

In the "bottoms-up" approach, a core concept in this process is the Total Addressable Market (TAM), which represents the total potential revenue opportunity if the product were to achieve full market penetration, assuming no competition or market constraints. In practice, however, that's rarely the case. A more realistic measure is the Serviceable Available Market (SAM). This looks at the portion of the total market that can actually be targeted based on factors such as patient eligibility (e.g., insurance coverage, adherence, etc.), competition, and anticipated market share. For our valuation model, we will focus on SAM, as it provides a more realistic estimate of peak revenue potential.

To see how SAM is calculated, let's walk through a high-level example using the framework shown in Figure 5 (Mauboussin and Callahan 2015). Although the framework is more detailed than we need right now, it highlights key considerations in market research, from estimating the potential patient pool to understanding competition, pricing, and patient behavior for forecasting drug adoption and sales.

COMPONENTS OF SERVICEABLE ADDRESSABLE MARKET

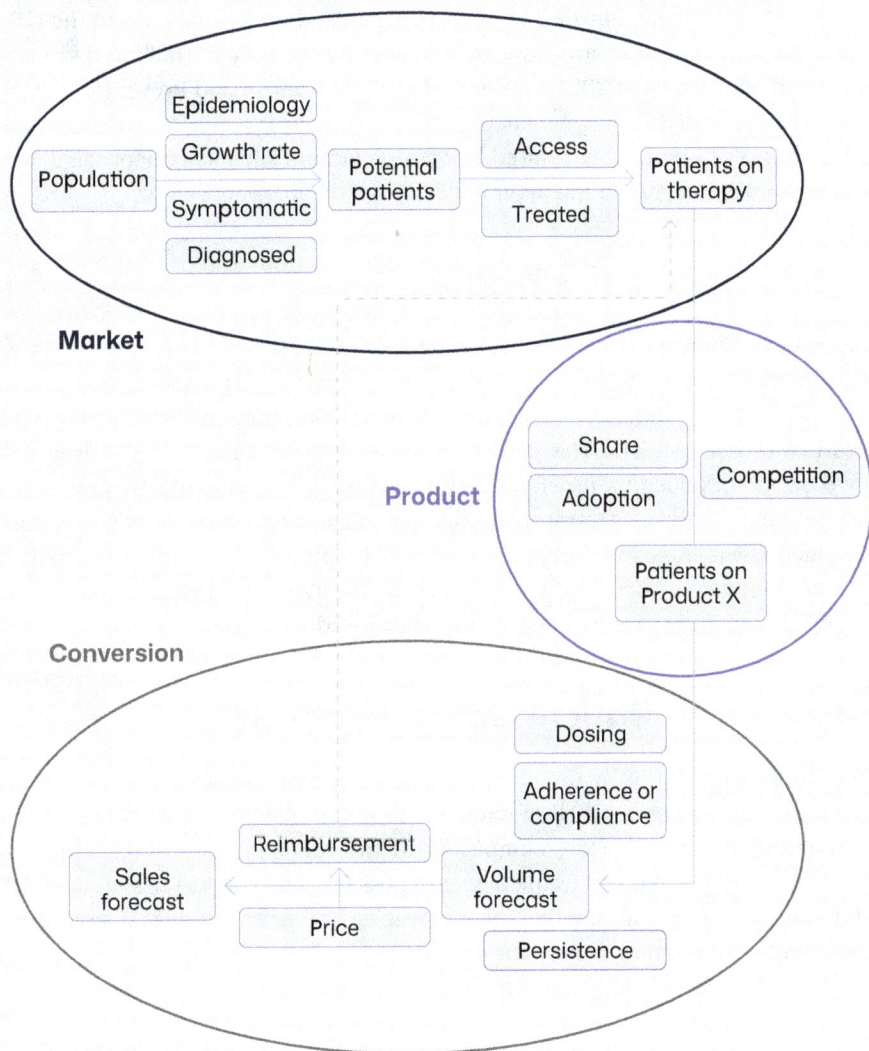

Adapted by AGMI Group from Mauboussin and Callahan 2015

Figure 5. *General components of SAM, including market, product, and conversion.*

The following breakdown examines the key components of SAM: Market, Product, and Conversion.

Estimating the Market Size

Imagine a biotech company developing a new medication for diabetes. In the U.S., around 38 million people are living with diabetes, but only 30 million had been officially diagnosed as of 2021 (Centers for Disease Control and Prevention 2024). However, not everyone with diabetes will be eligible for this new medication, and access could be further limited by various factors such as patient eligibility, insurance coverage, reimbursement policies, out-of-pocket cost, etc.

Understanding the Product

The company reports that the novel drug improves efficacy while having fewer side effects compared to existing options. There are various diabetes treatments available, including older drugs like metformin and glipizide in addition to newer ones like GLP-1 agonists. The market success of the theoretical new drug will hinge on factors such as its effectiveness, competition, adoption by physicians and patients, marketing strategies, and cannibalization (i.e., a new product introduced by a company reduces the sales of its previous products launched by the same company). Additionally, attributes like the method of delivery (oral versus injectable) and dosing frequency will influence its adoption rate.

Adjusting for Patient Conversion Rates

Even with a prescription, maintaining patient adherence can be difficult. Research shows that approximately 40% of patients stop their diabetes medications within six months, often due to the complexity of the regimen (based on method of administration, dosing, and frequency) and possible side effects (Liss et al. 2023). Additionally, factors like reimbursement policies and price will play a significant role in determining how widely the new drug is adopted, further affecting both initial uptake rates and long-term patient retention.

In summary, general considerations when using a bottom-up approach for market research include:

- consulting physicians or KOLs about unmet medical needs and their perspectives on current therapies
- understanding patient experiences with existing treatments
- identifying the product's likely therapeutic position (first-line versus second-line treatment)

- establishing an appropriate pricing strategy
- applying industry standards (or utilizing relevant comparators) to estimate product-specific lifecycle and market share

So, how does this concept apply to evaluating the market opportunity of a novel biotech product? For most early-stage biotech products, commercialization is still years off, meaning many critical details, such as regulatory timelines, competitive dynamics, and market adoption, remain uncertain. As a result, it's often impossible to determine every variable needed for a precise SAM calculation. Instead, we focus on three fundamental components to create a reasonable estimate:

1. **Total Addressable Patient Population**
2. **Market Share**
3. **Product Pricing**

In other words, the formula to calculate SAM is:

SAM = TAPP × MS × P

Where:

- SAM = Serviceable Addressable Market
- TAPP = Total Addressable Patient Population
- MS = Market Share
- P – Price

In the next few sections, we will walk you through these various components.

Sizing the Total Addressable Patient Population (TAPP)

In the previously discussed framework, we've outlined key factors that influence the SAM. A fundamental component of SAM is the TAPP—the total number of individuals a product can realistically target and serve. For the purpose of early-stage biotech valuation, we'll focus on a clear and practical approach for determining TAPP, also called a patient-flow forecast.

> **Quick tip:** When researching your target patient population, always rely on up-to-date and trustworthy sources like government health data, peer-reviewed medical journals, and reputable industry benchmarks.

Step 1

Determining the Number of Patients with a Particular Disease

When it comes to understanding how common a disease is, medical and epidemiological literature usually discusses two key terminologies: incidence and prevalence. These epidemiologic concepts help us understand not only how many new cases are emerging but also how many people are currently living with the disease.

- **Incidence** tells you how many new cases of a disease pop up over a specific time, often measured yearly.
  - Think of it as the number of people who just got sick. For example, if a report says the incidence of influenza is "50 cases per 1,000 people per year," it means 50 out of 1,000 people got the flu that year. Same goes for something like stroke, which might have an incidence of "150 cases per 100,000 people per year."

- **Prevalence** takes a bigger picture approach. It's like a snapshot that shows how many people are living with a disease at a given time, whether they were diagnosed last week or 5 years ago.

- For example, if the prevalence of influenza is "120 cases per 1,000 people," it means that at a point in time, 120 out of 1,000 people are currently living with the flu, which includes both the new cases and anyone still recovering from earlier infections. Similarly, if the prevalence of stroke is "500 cases per 100,000 people," it represents all individuals who have experienced a stroke, regardless of when it occurred.

While it might seem that incidence applies primarily to acute conditions and prevalence to chronic ones, this isn't always the case. Understanding the specific disease characteristics and applying clinical judgment is essential when determining which measure is more relevant for market evaluation.

Consider these examples as to why you cannot broadly apply these rules:

- **Deep vein thrombosis (DVT)**: While DVT is often an acute condition—where a blood clot forms suddenly and causes pain and swelling in the leg—some patients, such as those with cancer or recurring clots, may require long-term treatment with blood thinners (Arnold 2021). In those cases, prevalence might give you a better sense of the market since you're looking at everyone who needs ongoing care.

- **Gout**: This chronic condition causes sudden flare-ups of painful arthritis, often in joints like the knees. While managing gout over the long haul focuses on controlling uric acid levels, it's those sudden, acute gout attacks that need quick treatment (FitzGerald et al. 2020). For that reason, incidence may be the more useful measure when evaluating the market for fast-acting gout medications.

> **Quick tip:** It's always a good idea to consult with a medical expert, such as a healthcare provider, clinical researcher, or epidemiologist, to help determine which metric will be most useful!

Subdividing the Patient Population

Every patient is different, meaning not every patient with the same disease can receive identical treatment. For instance, acute stroke patients are segmented into those who are eligible for clot-busting therapy (i.e., tissue plasminogen activator) based on a small time window from the onset of neurological deficits. In general, the subdivision of the patient population is influenced by several factors, including the clinical judgment of the physician, comorbidities (additional medical conditions), demographics, type and severity of the disease, diagnostic imaging/ tests, and social factors (e.g., healthcare infrastructure, accessibility to treatment facilities, and insurance coverage).

Furthermore, while there may be established data for subdividing well-known diseases in financial models, many conditions require more nuanced clinical judgment. For example, how should patients with heart failure be subdivided? Should they be divided based on the heart's pathophysiology (e.g., heart pumping dynamics) or the symptomatology? In such cases, consulting with a physician or using data from a market research study can significantly enhance the credibility of your model, reflecting realistic treatment eligibility and improving its overall accuracy.

Once you have decided a specific patient population, there are two measurements that may be helpful when converting epidemiologic data into an addressable population.

Diagnosis Rate

The diagnosis rate is a crucial factor when determining how many people in a target population are correctly identified with a particular disease. Think about it: the development of new technologies (tests that are more accurate and have better sensitivity for disease detection), availability of screening tests, accessibility of healthcare, reimbursement rates, and even the training and awareness of healthcare providers all determine how effectively a population is diagnosed. Furthermore, diagnosis is influenced by symptomatic patients who seek healthcare to alleviate their symptoms.

There's a lot to consider here. For instance, healthcare access varies widely across regions. A community with advanced screening technologies and a highly trained workforce will identify more cases than one with limited resources. Similarly, clinical definitions and diagnostic criteria may shift, further complicating the diagnosis landscape. And let's not forget about practice norms/guidelines, regulations, and reimbursement policies that differ from one healthcare system to another, let alone different countries.

Consider these examples:

- **For asymptomatic conditions requiring primary prevention measures of such as type 2 diabetes or osteoporosis,** diagnosis rates depend on the use of screening tests in otherwise healthy individuals. Since these conditions often go unnoticed without symptoms, more frequent screenings, such as HbA1c tests (American Diabetes Association 2023) or bone density scans (White 2022), lead to higher diagnosis rates. This is influenced by factors such as insurance reimbursement, healthcare guidelines, and access to testing. When screenings are covered and recommended, more individuals get tested, increasing diagnoses. Conversely, limited coverage or access can result in underdiagnosis.

- **For late-presenting diseases such as pancreatic cancer,** which are typically diagnosed at an advanced stage due to subtle early symptoms (Pereira et al. 2020), early intervention is often limited by the absence of widely accessible and reliable screening tests. As a result, many patients are diagnosed when the disease has already progressed, reducing the chances for successful treatment. The development and regular use of effective early detection methods could significantly increase the number of patients eligible for curative therapies if the disease is caught at an early stage.

- **For diseases with evolving clinical definitions such as multiple myeloma,** advancements in risk stratification, genomic profiling, and biomarkers have significantly enhanced diagnostic accuracy (Visram et al. 2021). Growing awareness and continually updated diagnostic standards are essential for identifying high-risk patients early.

While many factors are involved, it's crucial to use your best judgment and clearly outline your assumptions when estimating diagnosis rates to determine the addressable population.

Treatment and Access Rate

The treatment rate represents the percentage of diagnosed patients within a target population who are eligible for treatment for a specific disease or condition based on physician and patient input. Not all patients who have been diagnosed with a disease may be appropriate for treatment with a given therapy. For many diseases, there may not be a clearly defined "gold standard" for care, and standards of care are constantly evolving as new research and therapies emerge. Consider these examples where treatment rates are lower than 100%, which should be factored into the valuation model:

- **For diseases with widely available nonmedical alternatives,** the number of patients eligible for medical treatment may be smaller. Take hypertension, for example—patients could opt for lifestyle changes such as diet modification, increased physical activity, or stress management instead of medication (Maniero et al. 2023). This means fewer patients might pursue the medical route since non-drug treatments are often preferred as a first-line option by both physicians and patients. The challenge lies in determining which patients would benefit most from medication versus other interventions, adding complexity to treatment planning.

- **For diseases with milder forms,** a significant fraction of patients may opt against drug therapy altogether. In cases like osteopenia or osteoporosis, patients often weigh the pros and cons, such as the outlook of the disease against the side effects, inconvenience, and risks of taking medication (Wang et al. 2017). Many patients opt for a "wait and see" approach or look for other alternatives. Understanding how patient choices, the burden of therapy, and disease progression interplay in these cases is crucial for accurately assessing the true treatment rate.

- **For diseases requiring last-resort interventions beyond standard care,** treatment rates are typically very low as patients must exhaust all conventional options before accessing advanced therapies. For example, experimental cell and gene therapies being developed for autoimmune conditions, such as systemic lupus erythematosus (SLE), would theoretically only be considered after patients have failed to respond to standard treatments such as corticosteroids, disease-modifying antirheumatic drugs (DMARDs), and biological immunomodulators. These cutting-edge interventions represent the final therapeutic option for the most treatment-refractory cases, where the severity of disease and failure of all previous treatments justify the significant costs and risks associated with advanced cellular or genetic modifications. The treated population in these scenarios is likely to be extremely small, typically representing less than 5–10% of patients diagnosed with the condition.

- **For early-stage diseases that rarely progress,** physicians often encounter the question of whether early intervention is the best approach. For example, with prostate cancer, there are early treatment options available, but since the risk of progression is often low and maybe even over-diagnosed, opinions on when to treat can vary (Leapman and Carroll 2017). Some patients may be offered active surveillance rather than immediate intervention. This means the treatment rate can fluctuate based on clinical guidelines, the healthcare provider's judgment, and the patient's personal preference. Flexibility in treatment models is essential here, especially as new research and data become available over time.

- **For diseases with evolving standards of care,** there is often no clear consensus on which approach is best. For instance, in idiopathic pulmonary fibrosis, some physicians may recommend antifibrotic medications to slow disease progression, while others might focus on managing symptoms through oxygen therapy or pulmonary rehabilitation (George et al. 2020). Given the unpredictable progression and varying patient responses to treatments, there is often no universally accepted approach, leaving physicians to tailor treatment plans based on individual patient factors.

These examples show how treatment standards—and patient uptake—can vary significantly depending on the context of the current clinical landscape of the disease. Healthcare providers often base their practice on current clinical guidelines, but when guidelines aren't clear or information is incomplete, their judgment and personal experience becomes even more important. This emphasizes the importance of including expert opinion and real-world clinical input to obtain accurate benchmarks.

Additionally, patients actively decide whether to pursue treatment based on factors such as disease severity, tolerance for potential side effects, and personal preferences, with some opting for alternative approaches such as lifestyle modifications or surgical interventions over traditional medical therapy. Completing patient interviews to understand their preferences and thoughts on new therapeutic approaches may also be helpful.

Access rate, distinct from treatment rate, refers to the percentage of eligible patients who can obtain a specific therapy when needed, accounting for all systemic barriers to treatment availability. This metric encompasses multiple factors that can prevent patients from accessing care, including geographic disparities where treatments may not be available in certain regions or healthcare systems, insurance coverage limitations that exclude or restrict access to specific therapies, distribution network constraints that limit where treatments can be obtained, and financial barriers created by high out-of-pocket costs or lack of reimbursement. Consequently, treatments with limited distribution channels, restrictive coverage policies, or prohibitive costs will have lower access rates, directly reducing the number of eligible patients who ultimately receive treatment.

> **Quick tip:** Obesity was once primarily managed through lifestyle changes, surgery, or other non-drug alternatives. However, this landscape has shifted dramatically with the introduction of Ozempic (semaglutide) for weight loss in 2021 (although it was originally approved for diabetes in 2017), which quickly became the top-selling drug across all pharmaceutical markets. This highlights the importance of staying current, as medical practices and treatment options can evolve very rapidly!

Step 3

Assessing Patient Adherence and Compliance

It's widely recognized that patients often face challenges with medication adherence. Some may forget to fill their prescriptions, while others might not follow dosing schedules as directed. A 2012 annual review on medication adherence highlighted a meta-analysis of 95 studies, which found that nearly 40% of patients discontinued their oral medications for various medical conditions by the end of the first year (Blaschke et al. 2012). This resulted in an adherence rate of approximately 60% after 1 year, as shown in Figure 6.

PATIENT ADHERENCE TO ORAL MEDICATIONS IN FIRST YEAR

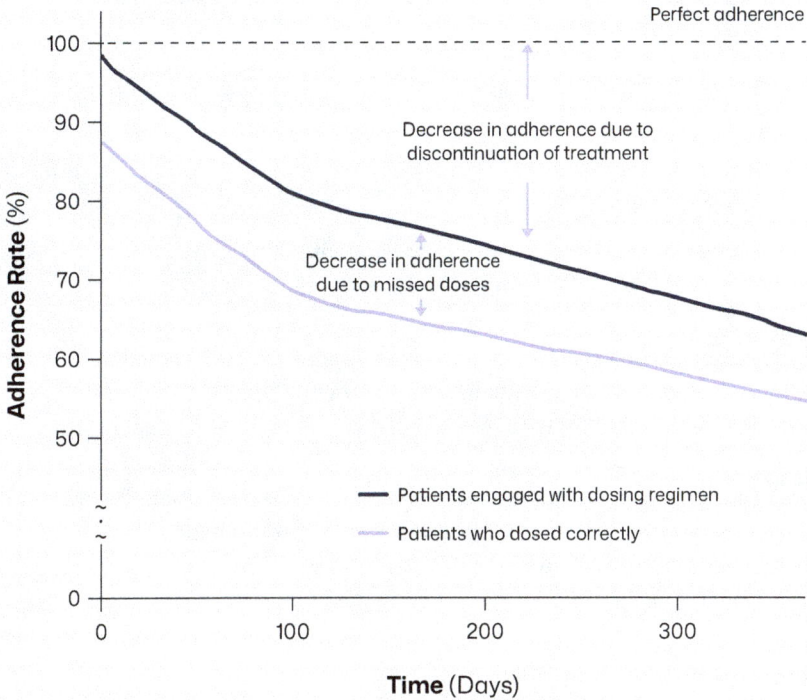

Source: Blaschke et al. 2012

Figure 6. Patient adherence to oral medications during the first year of treatment.

These findings are striking, and unfortunately, highlight a longstanding issue in the medical community: the gap between prescribed medications and actual medication adherence. A recent meta-analysis of 26 studies conducted between 2013 and 2023 revealed that global adherence rates for patients with type 2 diabetes taking oral antidiabetic medications were only 55% (Piragine et al. 2023).

From a biotech company's perspective, understanding how a medication is used in the real world requires factoring in how well patients follow their prescribed routines. Drug sales may fall short of expectations even when prescriptions are written due to three key factors:

1. **Filling the first prescription:** Not every patient picks up their initial prescription.

2. **Obtaining Refills:** Patients may begin treatment but fail to refill their prescriptions consistently or at all.

3. **Taking doses as prescribed:** Some patients skip doses, either intentionally or by accident, which can lower the expected refill rates.

While it might seem excessive to factor in adherence and compliance in early-stage biotech models, doing so helps prevent overly optimistic revenue forecasts and strengthens credibility with internal stakeholders and financial partners.

To model adherence accurately, it's not enough to consider patient behavior alone. We also adjust adherence and compliance rates based on external factors such as the treatment setting, disease type, drug formulation, route of administration, and whether the drug is branded or generic.

Another key consideration is whether the drug follows the 505(b)(2) pathway, a drug approval process that allows companies to submit new drug applications using existing data from previous studies or approved drugs to speed up the approval of similar treatments. Let's begin by looking at how treatment setting can impact medication adherence and compliance:

- **Outpatient Acute Care:** In acute outpatient care, the main concern is whether patients fill their prescriptions after receiving them. This is measured using Primary Adherence (PA), which can vary widely. For example, one study reported a 73% adherence rate for antibiotics generally (Park et al. 2018), while another found a 79% rate for bronchitis and urinary tract infections (Tamblyn et al. 2014). For modeling purposes, we generally assume an adherence rate of around 85% in acute settings. However, if you come across adherence rates above 90%, even in the most favorable cases, it's worth scrutinizing the study's credibility.

- **Outpatient Chronic Care:** For chronic conditions, it's important to assess whether patients consistently refill and take their prescriptions as prescribed. This is typically measured using the Medication Possession Ratio (MPR), which compares the amount of medication dispensed to what a patient should have taken over a given period, typically 1 year. In our valuation model, we assume an average adherence rate of around 70%, which aligns with findings from large European and Canadian cohort studies across various conditions (Blaschke et al. 2012; McNaughton et al. 2025). Keep in mind that MPR can vary significantly depending on the drug's formulation and the disease it treats—topics we'll discuss later in this chapter.

⅄ **Inpatient or Physician-Administered Care:** When medications are administered in an inpatient setting, adherence and compliance is assumed to be closer to 100%. However, for medications that require administration by a physician or healthcare provider in outpatient settings, adherence can vary between method of drug administration and severity of disease condition.

Other key factors that significantly influence a drug's adherence rate and overall value proposition include:

⅄ **Presence of Disease Symptoms**

- *Asymptomatic Diseases* (e.g., hypertension, hyperlipidemia): MPR tends to be lower, typically ranging from 50–70% (Birt et al. 2014; Tang et al. 2017).

- *Symptomatic Diseases* (e.g., gastric reflux): Patients may adjust their own dosing (self-titrate), leading to a reduced MPR (Hungin et al. 2012).

⅄ **Severity of Disease, Accumulated Pathological Conditions, and Perceived Medication Effectiveness**

- For severe conditions or accumulated pathological conditions (e.g., chronic myeloid leukemia, multiple sclerosis, amyotrophic lateral sclerosis), MPR is often higher, exceeding 75% and sometimes even around 85–90% (Haque et al. 2017; Kim et al. 2021; Higuera et al. 2016; Evans et al. 2021).

⅄ **Route of Drug Administration**

- Patients often report higher satisfaction with oral medications compared to infusions due to perceived efficacy and convenience (Hoffmann et al. 2024).

- However, infusions may result in better adherence than oral medications especially for those with accumulated pathological conditions such as multiple sclerosis, with rates approaching 90% (Engmann et al. 2021).

⋏ **Dosing Frequency**

- Medications with less frequent dosing schedules generally show higher adherence (Iglay et al. 2015).

⋏ **Combination Medications**

- Fixed-dose combination therapies tend to improve adherence and clinical outcomes compared to taking multiple separate medications (Elkout et al. 2012; Weisser et al. 2020).

⋏ **Drug Type (Branded versus Generic, 505(b)(2) Pathway):**

- Branded drugs may have higher adherence due to familiarity, promotion, and perceived quality, while generics might face lower adherence due to concerns over efficacy. However, generics offer a more cost-effective option (U.S. Food and Drug Administration 2023; Guy 2022).

- Although drugs approved through the 505(b)(2) pathway often enter the market faster and with lower drug development costs, this has not clearly shown to improve medication affordability or patient adherence.

An additional parameter of adherence/compliance is persistence, which measures the duration patients are on therapy. Obtaining the average treatment duration because it helps estimate the total number of treatment cycles a patient will complete before discontinuation, which directly impact revenue projections. When more detailed data is available, persistency curves provide deeper insights into patient retention patterns over time, showing exactly when and at what rate patients discontinue treatment. These curves are especially valuable for modeling how long patients remain on therapy after initial diagnosis or treatment initiation. However, when such granular persistency data is unavailable or of poor quality, using a simple duration metric such as "average time on therapy" provides a reasonable approximation for these financial modeling calculations.

Quick tip: Be sure to review the literature to identify if a specific indication is a significant outlier. For instance, MPR has been reported at or below 50% for antipsychotic treatments in patients with schizophrenia (Cho et al. 2022) and inhaled steroids in asthma patients (Jensen et al. 2021).

Avoiding Common Pitfalls

To ensure your TAPP estimate is accurate and reliable, keep the following considerations in mind to address potential weaknesses:

- **Misinterpretation of Epidemiological Data:** Be sure to correctly understand and differentiate between key terms such as incidence (new cases in a given time period) and prevalence (total cases at a specific point in time). Apply them appropriately to subdivide the patient population.

- **Poor Data Quality:** Use reliable, high-quality data that is representative of the population. For rare diseases, consider modeling different scenarios to account for data variability and uncertainty.

- **Ignoring Demographic Differences:** Ensure the data reflects the demographic characteristics of your target population, such as age, gender, education level, and ethnicity, as these factors can significantly impact outcomes.

By systematically addressing these four steps, you can more reliably size the addressable population—or TAPP. This is the first step to understanding the potential market size and, thus, a realistic estimate of the maximum revenue the treatment could generate.

> **Quick tip:** Factor in population growth when modeling future market size. The U.S. general population grows at approximately 1% annually (U.S. Census Bureau 2024), and disease populations typically grow proportionally. Incorporating this growth rate from the valuation date forward will provide more accurate projections for multi-year forecasts and help account for the expanding patient base over time.

Summary

Sizing Your TAPP

The process of determining your TAPP involves five key considerations:

01 Understanding Disease Prevalence/Incidence

Differentiate between prevalence (total existing cases) and incidence (new cases over time), selecting the appropriate metric based on disease characteristics and treatment goals.

02 Patient Population Subdivision

Recognize that not all diagnosed patients qualify for the same treatment due to factors like disease severity, comorbidities, and clinical guidelines. Consult medical experts to ensure realistic segmentation.

03 Accounting for Diagnosis Rates

Remember that diagnosis rates vary significantly based on screening availability, healthcare access, and evolving diagnostic criteria—all affecting your addressable market size.

04 Calculating Treatment and Access Rates

Understand that diagnosis doesn't guarantee treatment. Consider how treatment decisions are influenced by current medical treatment guidelines, provider recommendations, patient preferences, and alternative therapies. Consider how access to treatment is impacted by insurance coverage, out-of-pocket costs, geographic locations, and distribution networks.

05 Factoring in Adherence/Compliance

Incorporate realistic medication adherence/compliance rates based on treatment setting, administration route, dosing frequency, and disease severity.

We have included a comprehensive TAPP Estimation Checklist in the Resources section of this book to help you systematically work through each of these considerations when building your market model. This practical tool will guide you through the specific questions to address at each step, ensuring your projections are both data-driven and clinically realistic.

Estimating Market Share

Once you've estimated the TAPP, the next step is figuring out your market share—the percentage of total sales or revenue that your company or product captures in a specific market over a set period. It's important to be realistic; unless you are the first to enter the market (e.g., the first COVID-19 vaccine supplier), expect to share the space with competitors. Market share estimation is widely considered one of the most challenging aspects of biotech valuation because it requires predicting how the market will react to a new product and how that product will compete against future competitors.

Moving Beyond Using Analogues

A common approach is to simply look at historical "analogues"—similar products that launched previously—and assume your novel product will follow the same uptake pattern. This method has significant limitations because it assumes market conditions, competitive landscape, and company circumstances are identical to those of the historical product. In reality, numerous factors influence a product's market penetration that are rarely replicated exactly, including the launching company's reputation and resources, the strength of clinical data supporting the product, how well it addresses unmet medical needs, its differentiation from competitors, side effect profile, dosing convenience, the target physician specialty, patient population characteristics, and care setting requirements.

Rather than relying on a single analogue, a more robust approach involves using benchmarks derived from analyzing multiple product launches to identify consistent trends. By examining various comparable launches and deriving a distribution of performance outcomes, you can establish more reliable estimates that account for the variability inherent in product uptake. This statistical approach helps capture the uncertainty in market share projections while providing a more defensible foundation for your valuation assumptions.

Quick tip: When you do identify reliable analogue cases, use them to validate and refine your assumptions rather than as standalone predictors. This complementary approach strengthens your analysis alongside other inputs like industry benchmarks and expert interviews.

Identifying Competitors

For early-stage products, identifying competitors requires careful analysis of market focus, therapeutic areas, technology platforms, and product pipelines. Industry reports, online databases, and patent records can help you determine which companies are working on similar technologies or treating the same health conditions as your product.

When researching competitors, keep these guidelines in mind:

1. **Include Late-Stage Competitors**: If there's reason to believe a competitor will launch before your product loses exclusivity, you must include them in your model.

2. **Ignore Very Early-Stage or High-Risk Competitors**: If your competitors are in the very early stages or considered high-risk, it's okay to leave them out of your model.

3. **Use a Single Combined Competitor**: When there are multiple early-stage competitors, group them as a single combined competitor to account for the probability that one will launch before you, even if it's hard to say which one.

Quick tip: Within any market, the gold standard (typically the market leader) sets the benchmark against which physicians evaluate all competing products. By understanding what drives physician treatment decisions and how the current standard meets their needs, you can anticipate your product's ability to compete against gold standard products in the market and future new competitors.

Common pitfalls to avoid:

- ⋏ misidentifying comparable competitors
- ⋏ analyzing competitors too narrowly
- ⋏ overlooking competitor differentiation and technological strengths

By systematically considering these factors, you'll be able to identify the key competitors most relevant to your market share analysis and build more accurate projections for your valuation model.

How Does the Order of Entry Impact Market Share?

The timing of a drug's market entry can significantly affect its commercial success, especially in markets where products are largely undifferentiated (i.e., no single product offers a dramatic advantage over others). The term "me-too drug" or "follow-on drug" refers to a medication that is similar to a pre-existing drug, typically created by making minor modifications such as differences in side effect profiles or activity. In other words, such drugs are used to treat conditions for which medications already exist, usually without offering significant advantages over competitors.

First-in-class products and drugs typically enjoy substantial first-mover advantages by securing valuable lead time to build brand recognition, establish prescriber habits, define product labeling, and capture market share before competitors enter the market. This advantage translates into measurable market dominance: analysis of branded drug sales data from 2003 to 2014 revealed that first entrants maintained a 33-point market share increase over followers (Porath 2018). However, this advantage diminishes rapidly for subsequent entrants—second entrants showed only a 19-point market share increase over followers, third entrants dropped to 13 points, and fourth and later entrants plateaued at around 6 points. This pattern revealed that early followers can secure meaningful market positions, though the benefits decrease substantially with each successive entry.

Another analysis of 492 drug launches across 131 therapeutic classes between 1986 and 2012 found that first entrants achieve, on average, a 6-point market share advantage 10 years after first launch in the class compared to a "fair share" benchmark (defined as $1/n$ in an n-drug market) (Cha and Yu 2014). Notably, there is a greater advantage when there is a smaller market. For example, the market share advantage grows to 11 points in a two-drug market compared to just 4 points in markets with more than two drugs. In other words, in a two-drug market, the 1st entrant would have a 61% market share while the 2nd entrant would have a 39% market share. The analysis also revealed that while the 2nd to 5th entrants were not worse in market-share disadvantage (ranging from a market-share disadvantage of 1 to 3 points) compared to the fair share, there was a sharp decline for those ranked 6th and beyond (with a market-share disadvantage of 9 points).

Other impacts noted on market share advantage include route of administration (injection over oral meds), primary marketer (large pharma over other companies), medical specialization (specialty over primary), and prior experience (with experience versus no experience).

If first-move advantage is especially strong in a low differentiated market, first-in-class products can sometimes maintain over 60% of the market share, with later entrants securing less than 20%, as illustrated in Figure 7 for Lynparza (olaparib), Ibrance (palbociclib), and Januvia (sitagliptin) (Jacquet et al. 2024).

BRAND SHARE IN LOW DIFFERENTIATED MARKETS:
FIRST-IN-CLASS VS. LATE ENTRANTS

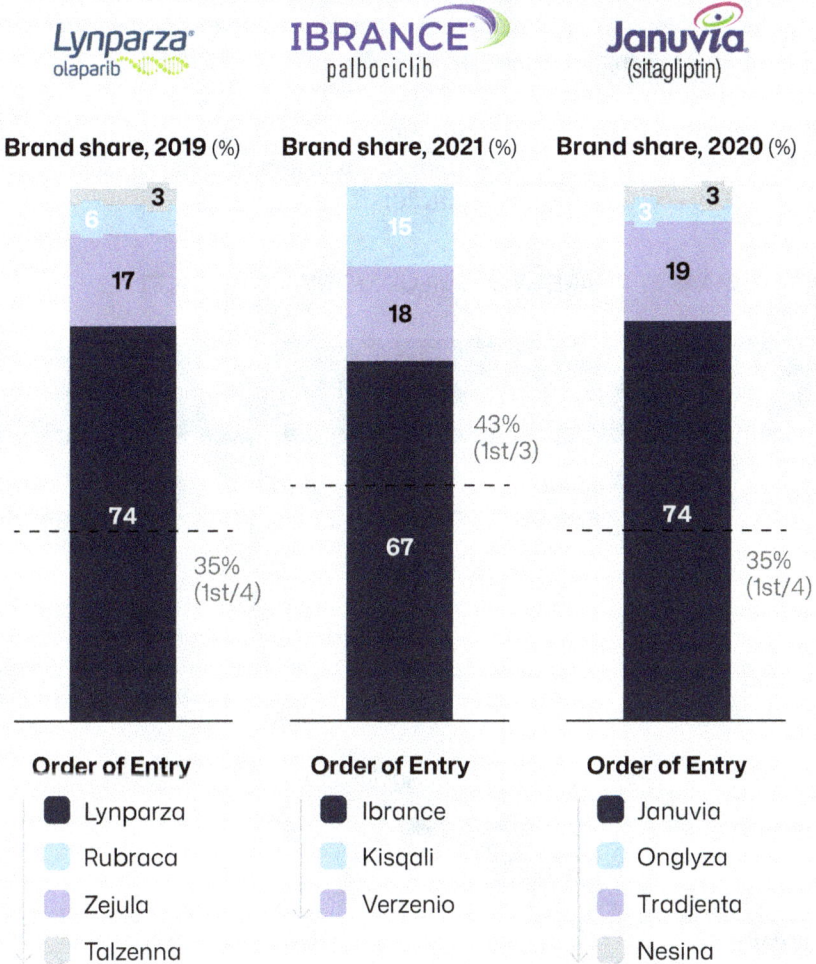

Lynparza®
olaparib

IBRANCE
palbociclib

Januvia
(sitagliptin)

Brand share, 2019 (%)   **Brand share, 2021** (%)   **Brand share, 2020** (%)

Lynparza: 3, 6, 17, 74 — 35% (1st/4)

Ibrance: 15, 18, 67 — 43% (1st/3)

Januvia: 3, 3, 19, 74 — 35% (1st/4)

Order of Entry
- Lynparza
- Rubraca
- Zejula
- Talzenna

Order of Entry
- Ibrance
- Kisqali
- Verzenio

Order of Entry
- Januvia
- Onglyza
- Tradjenta
- Nesina

– – – Order-of-entry benchmark (Nth player/Y-player market)

Note: Brand share is at steady state

Source: Jacquet et al. 2024

Figure 7.   Brand share distribution between first-in-class products and late market entrants in low-differentiated markets.

For most early-stage biotech valuation, we can approach this with a conservative estimate when we assume products are not very well differentiated. In these scenarios, the first entrant typically captures the largest share of the market due to its early presence and brand establishment, while subsequent entrants (up to four) divide the remaining share. As more competitors enter the market, the overall market share for each player tends to decline.

Table 3 presents one approach to projecting distribution of market share based on the order of entry into a market with mostly undifferentiated products based on the patterns of several analyses (Porath 2018; Cha and Yu 2014).

MARKET SHARE BY ORDER OF ENTRY AND MARKET SIZE

Market Size	Order of Entry				
	1st Entrant	2nd Entrant	3rd Entrant	4th Entrant	5th Entrant
Two-drug	61%	39%	-	-	-
Three-drug	38%	32%	30%	-	-
Four-drug	31%	25%	23%	21%	-
Five-drug	24%	21%	20%	18%	17%

Note: Market shares are estimated based on a market with low product differentiation.

Analysis by AGMI Group based on data from Porath 2018 and Cha and Yu 2014

Table 3.   Projected market share based on order of entry in a low differentiated market.

Note: We don't assume one product can capture 100% of the market. When a new biotech product becomes successful and profitable, other companies will develop competing products. By the time a product reaches its highest peak penetration in the market, there are almost always at least one or two other similar products available. Our calculations reflect this competitive reality by assuming a minimum of two products will be competing in the same market.

How about First-in-Class versus Best-in-Class?

As discussed earlier, being first-in-class usually captures a larger share of the market, even when later entrants are modestly undifferentiated and may have slightly better therapeutic benefits (Schulze and Ringel 2013). However, this is not always the case. For instance, Vumerity (diroximel fumarate), approved for multiple sclerosis, was able to outperform earlier entrants in certain segments despite not being first (Taylor 2022).

For products with significant differentiation or unique selling proposition, such as superior efficacy, safety, route of administration, or dosing frequency, the timing of market entry becomes less important. In these cases, a best-in-class product, even when launched years after the first-in-class, can sometimes capture the majority of the market share, as demonstrated in Figure 8 for Tagrisso (Osimertinib), Eliquis (apixaban), Firazyr (icatibant), and Fasenra (benralizumab) (Jacquet et al. 2024).

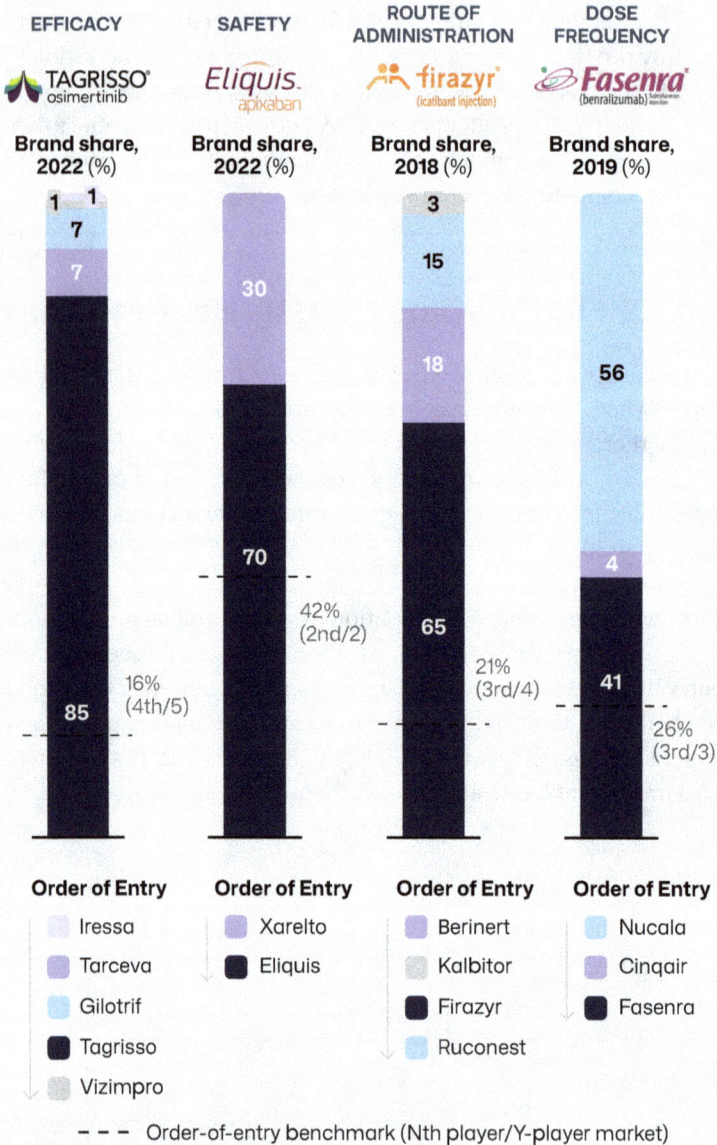

BRAND SHARE IN DIFFERENTIATED MARKETS: BEST-IN-CLASS LATE ENTRANTS VS. FIRST-IN-CLASS

Figure 8. Brand share distribution between best-in-class late entrants and first-in-class products in differentiated markets.

Note: Brand share is at steady state.

Source: Jacquet et al. 2024

A better way to compare first-in-class and best-in-class is illustrated in Figure 9, which shows a 2013 comparative analysis of various drugs between 2005 and 2010 by their order of entry and therapeutic advantage (considering factors like improved efficacy, side effect profiles, formulation improvements, etc.) (Schulze and Ringel 2013). The output quantifies the "value" captured by the drugs as a function of both order of entry and therapeutic advantage, measured by the net present value of their sales.

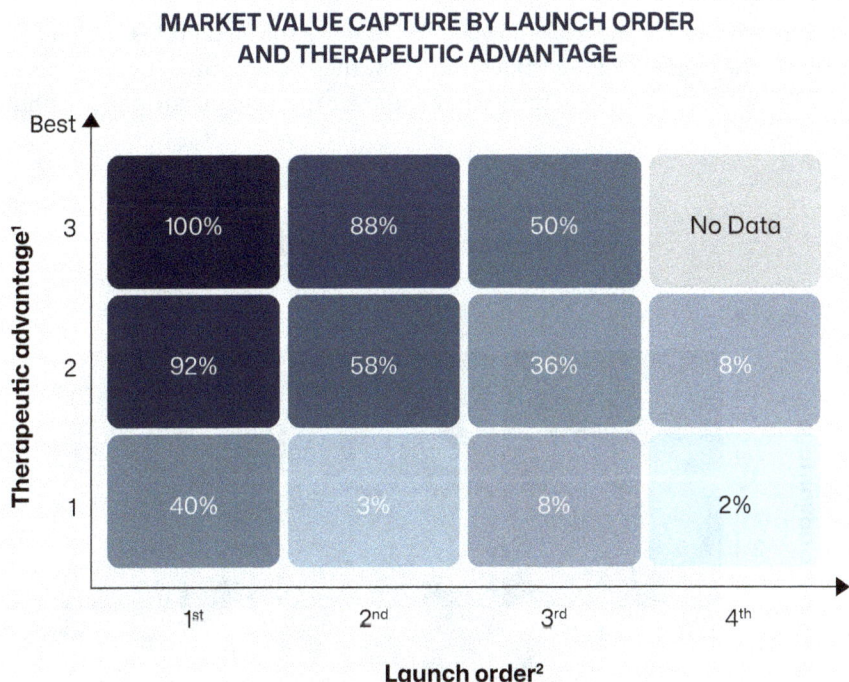

**MARKET VALUE CAPTURE BY LAUNCH ORDER
AND THERAPEUTIC ADVANTAGE**

Therapeutic advantage[1]	1st	2nd	3rd	4th
3	100%	88%	50%	No Data
2	92%	58%	36%	8%
1	40%	3%	8%	2%

Launch order[2]

[1] Compared to others in same mechanistic class
[2] Defined by date of approval by FDA

Source: Schulze and Ringle 2013

Figure 9. Market value captured as a function of order of entry and therapeutic advantage.

The bottom line from the 2013 analysis is that it is slightly better to be first than to be the best. Specifically, a first-in-class drug can retain up to 92% of its market value even when a later entrant is considered best-in-class, while a best-in-class drug launching second onwards can capture up to 88% of market value (Schulze and Ringel 2013).

However, being first to market is not sufficient. If a drug lacks significant differentiation, its market performance can suffer greatly as more competitors enter. Drugs with low distinctive value capture only 40% of the potential market even if they are first, and this can drop to less than 10% if other drugs have already entered the market. Additionally, the route of administration, which can impact convenience for the patient, can influence market adoption and long-term market share. For example, smoking cessation patches may compete differently against lozenges or gum due to patient preferences and convenience (Trigger et al. 2023).

If you are not the first, how fast do you need to move so you don't lose a significant portion of the market? Figure 10 illustrates how the value captured by new entrants varies based on the timing of their entry and their therapeutic advantage (Schulze and Ringel 2013).

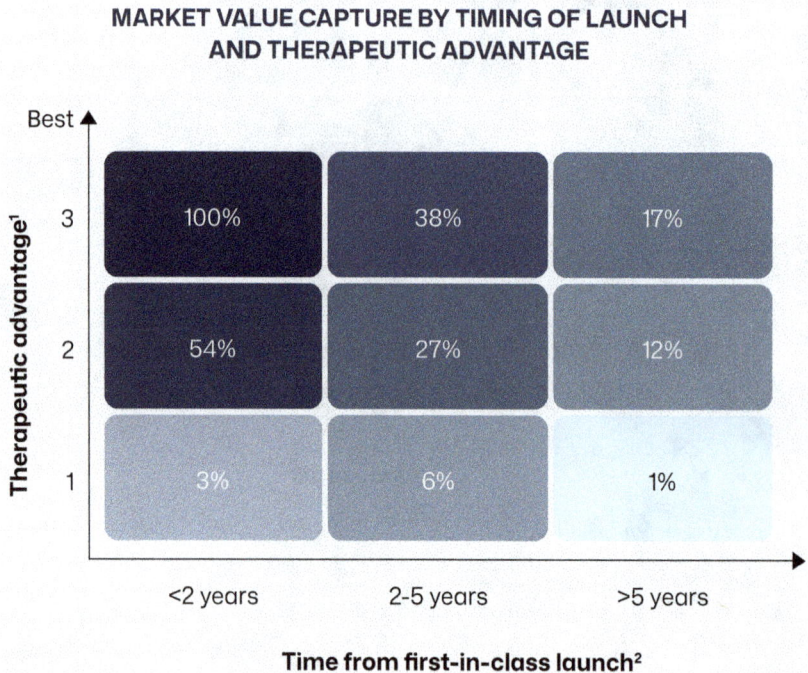

MARKET VALUE CAPTURE BY TIMING OF LAUNCH AND THERAPEUTIC ADVANTAGE

Therapeutic advantage[1]	<2 years	2-5 years	>5 years
3 (Best)	100%	38%	17%
2	54%	27%	12%
1	3%	6%	1%

Time from first-in-class launch[2]

[1] Compared to others in same mechanistic class
[2] Defined by date of approval by FDA

Source: Schulze and Ringel 2013

Figure 10. Market value captured by timing of first-in-class launch and therapeutic advantage.

For a best-in-class drug, entering the market within 2–5 years after the first-in-class typically captures about 38% of the value compared entering less than 2 years after the first-in-class. However, the longer you wait, the harder it becomes to gain traction. If you enter more than 5 years later, that value drops to just around 17% or lower. Timing is crucial, and the window for catching up is narrow.

As previously stated, order of entry and differentiation are the two key drivers of market success. However, there are three exceptions to consider:

1. **Targeted Therapeutic Advantage:** The precise nature of a drug's therapeutic advantage may shine in specific patient subgroups with unmet needs or has a breadth of indications that can broaden its market.

2. **Class Dynamics:** Some therapeutic classes are more welcoming to later entrants, particularly when drugs share the same mechanism of action and can be used as alternatives when initial treatments fail.

3. **Commercial Execution:** A highly effective commercial strategy such as a strong sales force can help carve out a niche in a crowded market.

Ultimately, estimating market share based on being differentiated or undifferentiated requires a nuanced, context-specific analysis. Defining meaningful differentiation goes beyond understanding the competitive landscape, it also involves considering factors such as disease progression, alternative treatment options, and patient needs. Without a clear and objective framework for assessing differentiation, key opportunities may be missed or misjudged, leading to suboptimal market strategies. Therefore, developing a comprehensive understanding of what truly sets a product apart from its competitors is essential for accurately predicting market share and commercial success.

Several other factors can impact market share, including a drug's efficacy, safety profile, dosing frequency, competitive pricing, and marketing strategies. These elements can provide later market entrants with a competitive advantage, potentially enabling them to capture a larger market share than initially projected.

Quick tip: Primary market research with physicians often overestimates market uptake due to two main sources of bias: inflation from the survey environment (since online surveys can't capture all real-world prescribing behavior) and inflation from respondent overstatement. This is particularly problematic for products with only modest differentiation where physician usage estimates are least reliable. There are several methods to address this bias: applying a standard research "haircut" (such as 50% off physician responses) or using top-two box scores (combining the highest two positive responses from rating scales) (Aggarwal et al. 2017; Fram and Gogolina 2023). However, arbitrary standard haircuts might not be appropriate for every market study, as each new therapy has multiple factors to consider. A more structured adjustment approach should account for: (1) payer access restrictions, (2) patient fill rates, and (3) commercial reach limitations. Additionally, use allocation-based questioning that forces physicians to consider tradeoffs between all treatment options within specific patient segments rather than estimating uptake in isolation. For products at parity with competitors, expand analysis beyond clinical data to examine competitive pressures, unmet need, and patient engagement factors. This methodical approach may provide more nuanced and accurate demand forecasting.

Summary

Estimating Your Market Share

The process of estimating your product's market share involves four key considerations:

Competitive Landscape Analysis

01 Identify relevant competitors based on their development stage, including products likely to launch before your product. Consider grouping early-stage competitors into a single combined competitor to account for future market entry without overcomplicating your model.

Market Entry Timing Assessment

02

Recognize that order of entry significantly impacts market share, with first-in-class products typically maintaining a 6% advantage over "fair share."

Product Differentiation Evaluation

03

Determine whether your product will be modestly undifferentiated or significantly differentiated from competitors. For truly differentiated products with superior efficacy, safety, or convenience, even later market entrants can capture majority market share, as demonstrated by products like Tagrisso and Eliquis.

Strategic Positioning Factors

04

Consider special circumstances that may influence market share beyond the standard models, including targeted advantages in specific patient subgroups, therapeutic class dynamics that welcome later entrants, and commercial execution strategies that can carve out market niches.

We have included a comprehensive Market Share Estimation Checklist in the Resources section of this book to help you systematically work through each of these considerations when building your market model. This practical tool will guide you through the specific questions to address at each step, ensuring your projections reflect competitive realities and avoid overestimation pitfalls common in early-stage biotech valuations.

Estimating Drug Price

Now that we've covered how to estimate market share, let's dive into pricing, which is equally crucial for driving revenue.

Key Pricing Definitions and Distribution Dynamics

Drug pricing in the biotech industry involves complex terminology and distribution relationships that are essential to understand for accurate valuation. Many terms you'll encounter aren't actual transaction prices but rather pricing metrics or reference points representing different stages of the pricing process and various stakeholder perspectives.

Here are the key pricing definitions you should be familiar with (from lowest to highest dollar amount) as well as general resources to find these drug prices:

1. **Average Sales Price (ASP):** The "sales price" defined by federal law that is paid to manufacturers by all purchasers, net of all rebates and discounts

    a. Freely available for a subset of physician-administered drugs covered under Medicare Part B on the Center for Medicare and Medicaid Services as ASP Pricing Files (Centers for Medicare & Medicaid Services 2025)

    b. **Key model input:** Include this in your financial model as it more closely estimates what the manufacturer actually receives

2. **Wholesale Acquisition Cost (WAC):** The "list price" defined by federal law at which manufacturers sell drugs to wholesalers or direct purchasers, which does not include any discounts or rebates

    a. For single-source, branded pharmaceuticals, prices paid by retail pharmacies in the Federal National Average Drug Acquisition Cost (NADAC) database closely mirror WAC (Medicaid 2025; Bruen and Young 2014)

    b. Prices calculated and published by Medi-Span (Wolters Kluwer 2025) and MicroMedex "Red Book" (paid tool for drug pricing, data, and manufacturer information) (Micromedex Solutions 2025)

3. **Retail Price (RP):** The average price charged to the patient or end user

    a. For outpatient drugs, refer to prices from pharmacy chains such as CVS, Walgreens, Amazon Pharmacy, Walmart, or OptumRx

4. **Average Wholesale Price (AWP):** Estimate of the average price retail pharmacies pay to wholesale distributors

    a. Refer to prices published in Federal NADAC database (Medicaid 2025), Medi-Span (Wolters Kluwer 2024), First Databank (First Databank 2025), and MicroMedex "Red Book" (Micromedex Solutions 2025)

Quick tip: AWP serves as a benchmark pricing metric but is intentionally set above real transaction prices. This inflation explains why retail prices are often lower than published AWP figures—pharmacies typically pay wholesalers less than AWP through privately negotiated discounts, and Pharmacy Benefit Managers (PBMs) help pharmacies negotiate lower costs with manufacturers that get passed to retail pricing.

Two critical distinctions often confuse industry newcomers: the difference between wholesalers and Group Purchasing Organizations (GPOs), and their respective roles in the pricing chain. Wholesalers directly purchase pharmaceutical products from manufacturers, maintain inventories, and handle physical distribution to pharmacies and healthcare providers. GPOs, conversely, don't purchase or distribute drugs—they act as collective negotiators on behalf of healthcare providers to secure favorable pricing and contract terms.

PBMs add another layer of complexity by operating between manufacturers, pharmacies, and third-party payers. PBMs manage prescription drug benefits for health plans, negotiate rebates with manufacturers, and establish formularies that determine coverage and patient costs. They essentially act as intermediaries that influence both what payers reimburse and what patients pay at the pharmacy counter.

These intermediaries significantly influence pricing at each stage of the supply chain, from manufacturer through wholesaler to pharmacy, with PBMs uniquely positioned to affect both upstream manufacturer negotiations and downstream patient access.

Understanding these distinct roles and their interrelationships is essential for accurate revenue modeling. Misunderstanding can lead to flawed pricing strategies, inappropriate partnership expectations, or inaccurate market assumptions that undermine commercialization efforts.

Figure 11 illustrates these key pricing terms and stakeholder relationships within the drug distribution ecosystem (Prescription Analytics 2025). Note that third-party payers are not depicted in this figure.

DRUG PRICING AND DISTRIBUTION CHANNEL DYNAMICS

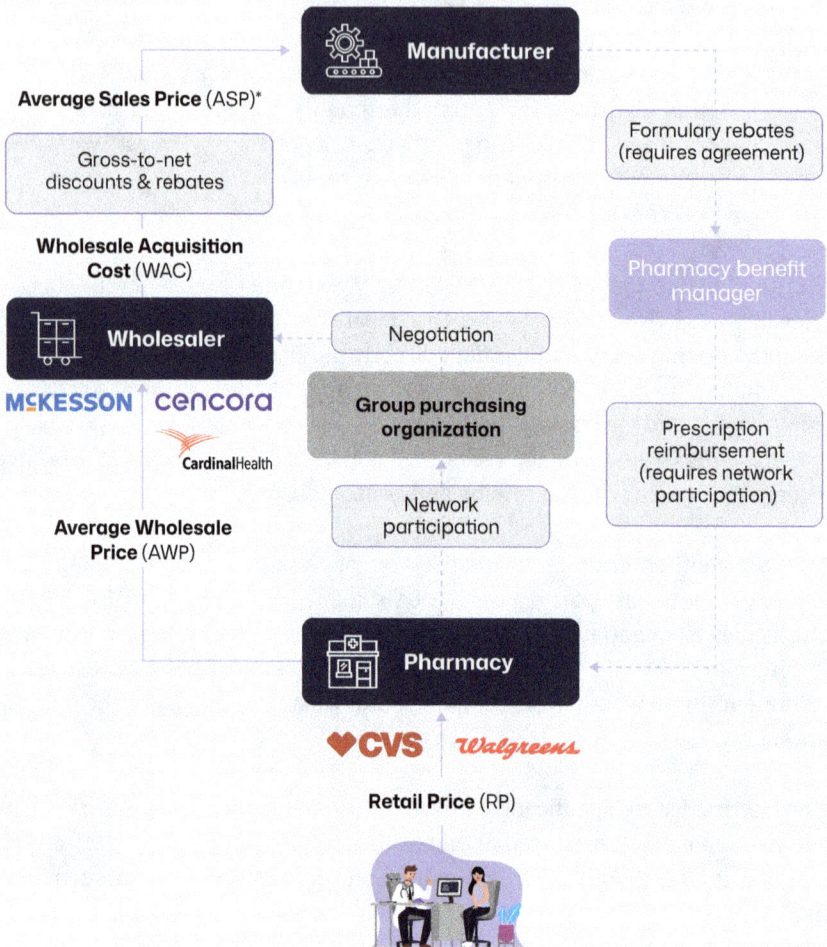

Figure 11. *Overview of drug pricing terms and how key distribution entities influence them.*

* Average sales price can be used as a net price proxy in relation to gross-to-net discounts and rebates.

Note: Third-party payers and health plans are not depicted in this figure.

Adapted by AGMI Group from Prescription Analytics 2025

Pricing for biotech products or drugs can vary depending on the source. To help you navigate these differences, we've provided you a general reference in Table 4 that converts between various pricing variables for branded drugs, based on our industry benchmarks and analysis (IQVIA 2023; Schumacher and Miller 2021; MedPAC 2019; Fein 2012; Curtiss 2010; U.S. Department of Health and Human Services 2005). Note that these conversion rates represent general rules based on averages and may vary significantly in practice. For instance, the ASP is estimated around 84% of the WAC, but this conversion can vary widely—ranging from 70% to 90% depending on the drug and market conditions (IQVIA 2023; MedPAC 2019).

DRUG PRICE CONVERSION TABLE

		(2) Choose Price to Convert to		
		AWP	**Retail**	**WAC**	**ASP**
(1) Select Reference Point	AWP = 100	100	93	83	70
	Retail = 100	107	100	90	75
	WAC = 100	120	112	100	84
	ASP = 100	143	126	113	100

Note: These figures are approximations intended for general reference. For precise data, please refer to updated industry price databases.

Analysis by AGMI Group based on data from IQVIA 2023, Schumacher and Miller 2021, MedPAC 2019, Fein 2012, Curtiss 2010, and U.S. Department of Health and Human Services 2005

Table 4. *Reference price conversion table for key pricing metrics. To use: (1) select your reference point, then (2) choose the price type to convert to.*

Now that you have an understanding of key drug pricing definitions and where to find them, let's explore how these pricing figures relate to each other in practice. Specifically, how do we move from a drug's list price, aka the WAC, to the actual price that manufacturers receive?

This brings us to an essential concept in biotech and pharma pricing: gross-to-net.

Introduction to Gross-to-Net: What are deductions?

In the biotech or pharma world, there's a concept called "gross-to-net." This term describes the gap between a drug's list price—typically the WAC—and the net price that manufacturers actually receive after accounting for various mandatory and strategic deductions. These deductions encompass a wide range of discounts and rebates that pharmaceutical companies must provide to different stakeholders throughout the healthcare ecosystem. In short, common deductions include:

- **GPO discounts** for bulk purchasing agreements
- **Medicaid and Medicare rebates** mandated by government programs
- **commercial payer rebates** negotiated with insurance companies
- **co-pay assistance programs** that reduce patient out-of-pocket costs
- **product returns** from distributors and pharmacies
- **chargebacks** from indirect purchasers receiving contract pricing

Manufacturers implement these deductions to secure formulary placement with payers, comply with government program requirements, improve patient access, or maintain competitive positioning in the market. These deductions are typically made to incentivize certain purchasing behaviors or meet contractual obligations, which ultimately affect the manufacturer's realized revenue. Note that certain expenses like GPOs and PBMs administrative fees are operational costs that biotech and pharma companies pay but are calculated differently from the primary revenue deductions.

Figure 12 visually depicts the funnel of discounts, rebates, and fees that impact the gross-to-net sales.

GROSS-TO-NET REDUCTION FUNNEL

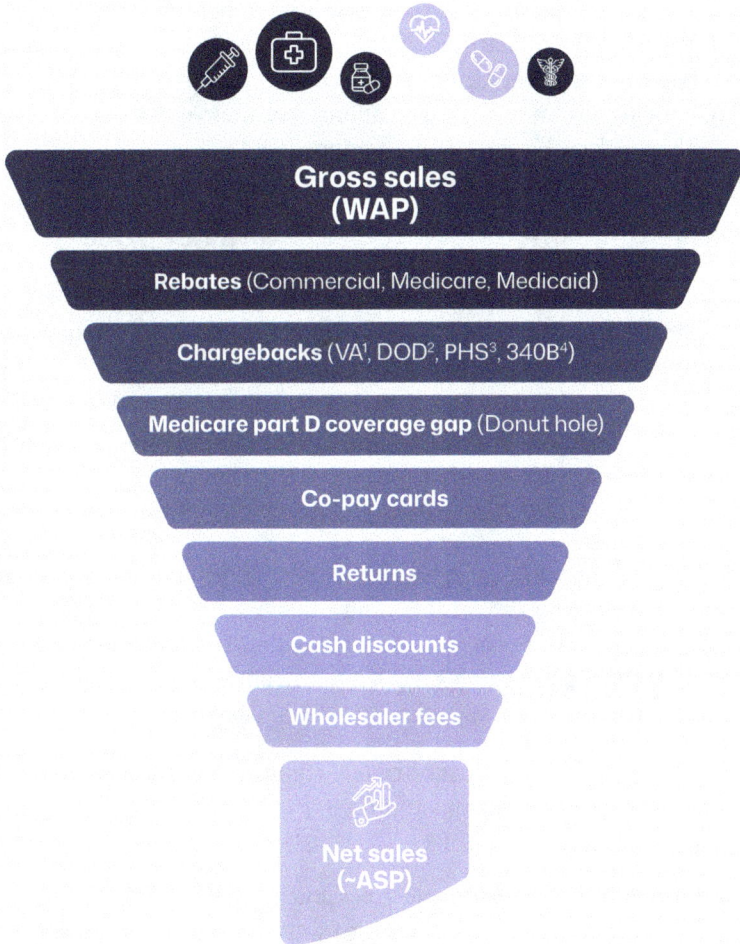

Gross sales (WAP)

Rebates (Commercial, Medicare, Medicaid)

Chargebacks (VA[1], DOD[2], PHS[3], 340B[4])

Medicare part D coverage gap (Donut hole)

Co-pay cards

Returns

Cash discounts

Wholesaler fees

Net sales (~ASP)

[1] Veterans Affairs department
[2] Department of Defense
[3] Public Health Service
[4] 340B Drug Pricing Program

Figure 12. Funnel diagram showing the progression of gross-to-net price reductions.

Calculating the gross-to-net adjustment for individual biotech products is challenging, as it depends on policies and negotiations among various entities, such as the government, pharmacies, PBMs, GPOs, and insurers, many of which are commercially sensitive and are not publicly disclosed. In addition, manufacturers are generally keen to keep their gross-to-net details private, as these margins can reveal the inner workings of their payer and GPO contracting strategies. Additionally, as a therapeutic area becomes more competitive with more products entering the market, manufacturers often face increased pressure to cede margin through larger discounts and rebates. For a more detailed discussion on the key factors influencing gross-to-net calculations and their impact on a biotech or pharma product's lifecycle, please refer to Appendix 1.

What's the difference between ASP and net price?

ASP is a distinct regulatory metric used by the Centers for Medicare & Medicaid Services (CMS) to determine Medicare reimbursement rates. It is not synonymous with net price, though it can sometimes serve as a proxy for net pricing in financial models. ASP represents a specific definition and calculation mandated by CMS and may not fully reflect a manufacturer's actual net revenue realization. ASP figures are publicly available through CMS ASP Drug Pricing Files, which are updated quarterly.

Net price, by contrast, represents the actual price a manufacturer receives after accounting for all discounts, rebates, and other concessions across all payers and channels. As noted previously, this information is considered commercially sensitive and confidential, making it challenging to obtain precise figures. However, net price and gross-to-net deductions can be estimated using several sources:

- **SSR Health net price database** is a commercial database with net price estimates based on manufacturer-reported net revenue and volume data (SSR Health 2025).

- **Company 10-K filings** provide gross and net sales data to calculate gross-to-net discount rates (U.S. Securities and Exchange Commission 2025).

- **Investor presentations** often contain gross-to-net metrics.

- **State drug cost transparency reports** are required in certain states such as California and Maine (California Department of Health Care Access and Information 2025; Maine Health Data Organization 2025).

There are three methods to estimate net price:

1. **WAC-Based Calculation**

   a. Get WAC from reliable source (e.g., NADAC, Red Book, Medi-Span)

   b. Estimate gross-to-net discount rate using SSR Health database, 10-K filings, or other industry reports

   c. Apply formula:

   Net Price = WAC × (1 - Gross-to-Net %)

2. **ASP as Proxy**

   a. Use CMS provided ASP Pricing Files as net price estimate, which is best used for Medicare-focused analysis or products with significant Medicare populations

3. **Published Net Price Databases**

   a. Access net pricing tool or database (e.g., SSR Health database)

While these estimation methods provide starting points for net price analysis, let's take a look at what market pressures drive gross-to-net deductions.

Gross-to-Net Dynamics in a Modern Era

In general, most of these price reductions come from rebates that drug manufacturing companies pay to third-party payers and U.S. government programs, as shown in Figure 13 of the total value of gross-to-net reductions of manufacturers in 2021 (Fein 2022).

MANUFACTURER GROSS-TO-NET REDUCTIONS FOR BRAND-NAME DRUGS, 2021

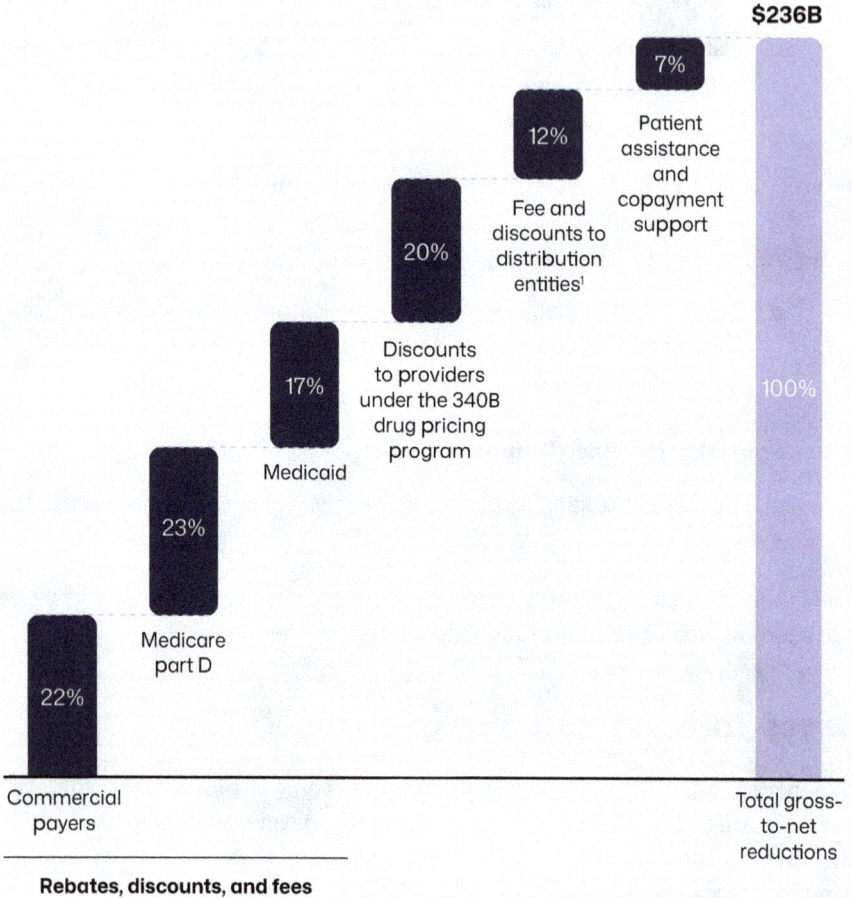

$236B

7%

Patient assistance and copayment support

12%

Fee and discounts to distribution entities[1]

20%

Discounts to providers under the 340B drug pricing program

17%

Medicaid

23%

Medicare part D

22%

100%

Commercial payers

Total gross-to-net reductions

Rebates, discounts, and fees

[1]*Payments by manufacturers include: administrative fees to PBMs; fees and discounts to pharmacies and wholesalers; and all other off-invoice discounts and rebates*

Source: Fein 2022

Figure 13. Manufacturer gross-to-net reductions for brand-name drugs in 2021.

It's important to recognize that discounts from list to sales price can differ significantly based on company partnerships, drug market, and whether the drug is branded or generic. For instance, in 2023, major pharmaceutical companies offered average discounts ranging from 37% to 74% on their brand-name drug portfolios (Fein 2024a). Consequently, it's important to analyze sales price in relation to list price or other pricing metrics, particularly for drug categories experiencing strong payer pressure, such as primary care markets with numerous competitors, or for outpatient medications provided to providers in the 340B drug pricing program, a federal program that requires drug manufacturers to offer steep discounts to hospitals and clinics that serve a high proportion of low-income or uninsured patients.

Meanwhile, more than half of these reductions came from payments to PBMs and other third-party payers, while discounts from 340B program and Medicaid rebates accounted for another third (Fein 2024b). This underscores the growing influence of PBMs, whose industry remains highly consolidated, further strengthened by the rise of group purchasing organizations that negotiate rebates on behalf of their members. Additionally, the Inflation Reduction Act of 2022 will likely further discourage manufacturers from raising list prices on brand-name drugs with significant Medicare utilization, whether patient- or provider-administered. Therefore, it would be reasonable to discount sales price further relative to list price or other pricing measures.

With mounting pressure from growing gross-to-net reductions over the past decade or so, manufacturers have continued raising list prices to sustain revenue, further widening the gross-to-net gap. Figure 14 shows the increasing total value of gross-to-net reductions for brand name drugs from 2019 to 2023, which has grown by approximately 10% year over year (Fein 2024a).

TOTAL MANUFACTURER GROSS-TO-NET REDUCTIONS OF BRAND-NAME DRUGS, 2019 TO 2023

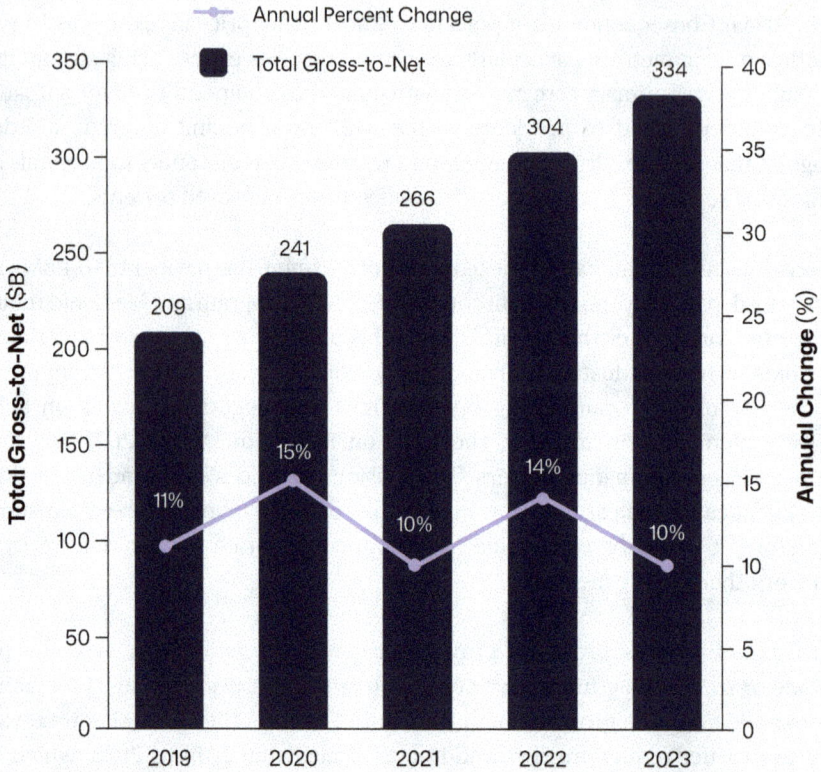

Source: Fein 2024a

Figure 14. Total Manufacturer gross-to-net reductions of brand-name drugs from 2019 to 2023.

Our key takeaway is that gross-to-net considerations will be critical in shaping a biotech company's pricing strategy for their products, as gross-to-net directly impacts the revenue received. With PBMs and other third-party payers exerting increasing influence through rebate negotiations, biotech companies must carefully assess the financial impact of various pricing approaches while strategically maintaining confidentiality around their actual gross-to-net margins. This opacity helps protect their competitive positioning and negotiating leverage.

Therefore, pricing strategies will be key in balancing competitive market access with sustainable revenue generation, factoring in payer dynamics, reimbursement trends, and evolving policy changes.

Framework for Calculating Drug Pricing

Now that we've covered the various pricing definitions and the impacts of gross-to-net, let's walk through the general steps for estimating drug prices, particularly for the valuation of an early-stage biotech product:

1. **Model a Starting Price in the U.S. Market:**
   - Estimate the drug's market entry price by identifying comparable therapies and referencing ASP values, converting other pricing benchmarks to ASP where needed, using gross-to-net adjustments, or consulting net price databases.
   - Convert per-unit dosing into annual revenue per patient by factoring in dosage, frequency, duration, and indication.

2. **Account for Annual Price Adjustments:**
   - After the launch year, adjust pricing annually based on whether the market is located in the U.S. or internationally

3. **Estimate International Prices:**
   - Apply a correction factor to the U.S. launch price to approximate global pricing.
   - Adjust for country- or region-specific price regulations and expected annual increases.

> **Quick tip:** Because valuation models are highly sensitive to pricing, it is important to be methodical, transparent, and conservative to ensure accurate estimates.

Annual U.S. price increases are typically modeled at around the current U.S. inflation rate, for example, 2–3% for 2024 to 2025 (U.S. Inflation Calculator 2025) starting from launch year, reflecting general inflation trends and historical estimates of net price growth. However, this pattern has been somewhat disrupted in the post-COVID-19 era with periods of sharp increases and decreases in inflation rate

(IQVIA 2024). For more precise price increase amounts and cadence, reference the Red Book list price databases, which provide detailed historical pricing increases for specific drugs.

> **Quick tip:** When using pricing data from sources like equity analysts, consultants, government reports, or company earnings releases, carefully evaluate both the context and reliability of the information. Be mindful of the specific pricing metric used (e.g., ASP versus retail price) to ensure accuracy. Avoid incorporating overly aggressive price premiums, as this can undermine the credibility of your model with investors.

International Pricing Adjustments

Drug prices are substantially lower outside the U.S. in most major markets due to several key factors. These include value-based pricing, external reference pricing that excludes high U.S. prices, direct government negotiation with pharmaceutical companies, and stricter price controls. Additionally, many countries focus more on promoting generic drugs and have tighter regulations on drug advertising.

Given these significant differences, when estimating international pricing for pharmaceuticals, it's crucial to adjust both the initial prices and the projected annual increases. To accurately model international drug pricing, we can employ two separate approaches:

1. **Adjusting U.S. ASP with a Conversion Factor**

    - In major markets like the EU-5 (Germany, Spain, France, Italy, and the UK) and Japan, gross drug prices average 36% of U.S. prices, seen in Figure 15 (Mulcahy et al. 2024). The difference is even more pronounced for branded drugs, where prices average 24% of U.S. levels.

    - For other European countries and global markets, pricing assumptions typically align with the EU-5.

2. **Using Direct Price Comparisons from International Markets**

    - International price data can be sourced from organizations like Health Action International, which aggregates national drug prices across various countries (Health Action International 2019).

- However, many reports focus on retail prices, which are often higher than the ASP. Without access to country-specific retail-to-manufacturer discount rates, adjusting this data accurately can be challenging (Perehudoff et al. 2021).

**GLOBAL GROSS PRESCRIPTION DRUG PRICES
BENCHMARKED AGAINST THE U.S., 2022**

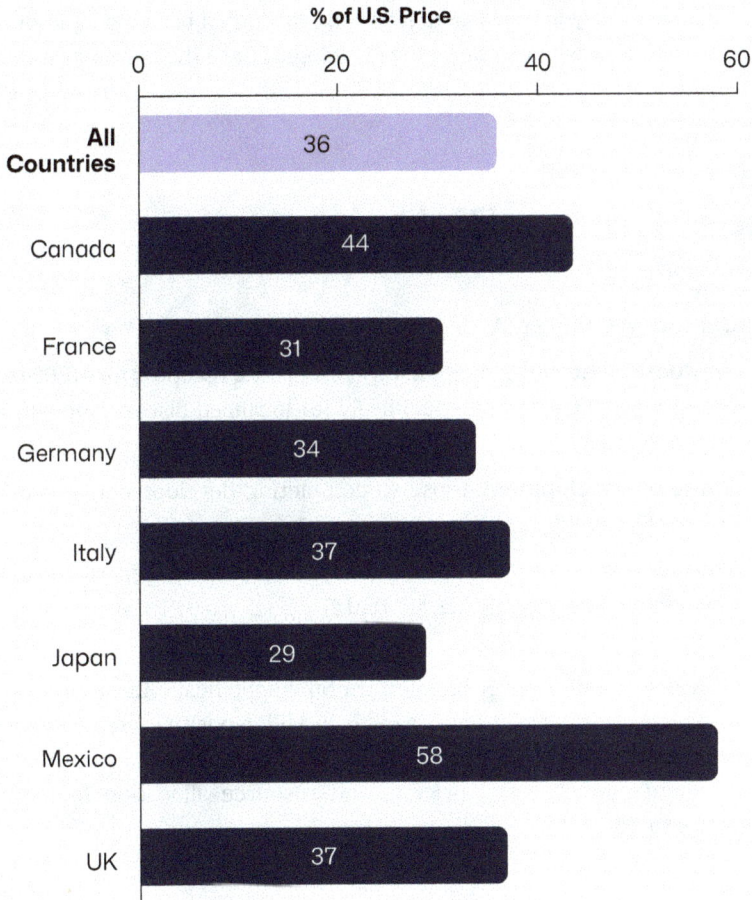

% of U.S. Price

Country	% of U.S. Price
All Countries	36
Canada	44
France	31
Germany	34
Italy	37
Japan	29
Mexico	58
UK	37

Source: Mulcahy et al. 2024

Figure 15. Comparison of gross prescription drug prices across international markets using U.S. pricing as the baseline benchmark for 2022.

Annual international price increases are generally assumed to be either flat or slightly negative year-over-year price changes in valuation models (U.K. Department of Health and Social Care 2018; Financial Times 2023; Trinity Life Sciences 2023).

That was a lot to take in—but you now have a clear framework for navigating drug pricing and applying it to early-stage biotech valuation.

To help solidify what you've learned, let's walk through a brief example of market analysis for a HER2-positive breast cancer biologic to see these concepts in action.

CASE STUDY
Breast Cancer Biologic

- **Company:** Company A
- **Product/Asset:** Novel biologic targeting HER2 receptor for HER2-positive breast cancer treatment, specifically for localized disease population (i.e., no metastasis)
- **Stage of Development:** Close to completing development, preparing for U.S. market entry
- **Valuation Method:** Market sizing analysis bottom-up approach to calculate Serviceable Addressable Market (SAM)

Company A developed a novel biologic therapy specifically targeting the HER2 receptor, a critical driver of tumor growth in HER2-positive breast cancer. The company needed to estimate potential U.S. market revenue to inform strategic decisions about market entry, pricing, and resource allocation for this novel oncology treatment.

BREAST CANCER THERAPEUTIC EXAMPLE: SAM CALCULATION METHODOLOGY

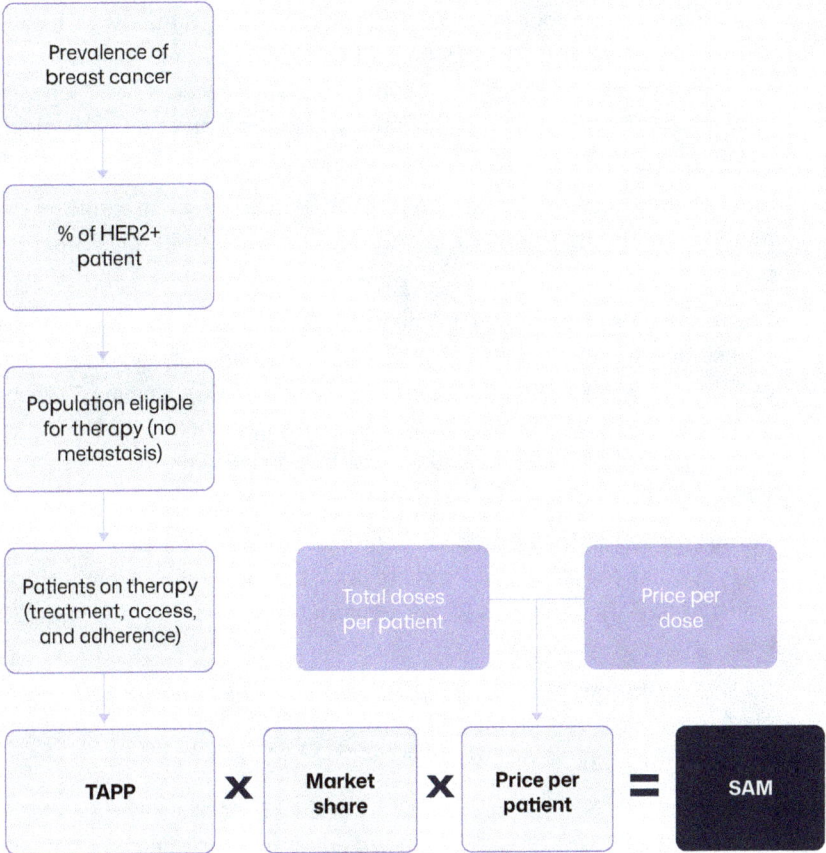

Figure 16. Flowchart showing simple market analysis for a novel breast cancer biologic example, depicting the elements to calculate SAM.

Key Assumptions

- ⌃ **Market Size:** Total U.S. breast cancer prevalence of around 4 million women in 2022 (SEER 2025), with 14% being HER2-positive (560,000 patients) (American Cancer Society 2022), further narrowed to 66% with localized disease (369,600 eligible patients) (SEER 2025). Assume 90% treatment rate (Ellegård et al. 2022), 60% national coverage or access rate (Genentech 2023), and 85% adherence/compliance rate (Engmann et al. 2021), yielding 169,646 patients receiving therapy

- ⌃ **Market Penetration:** Assume max of 20% market share capture given multiple competitors already in the market resulting in 33,929 potential patients

- ⌃ **Treatment Protocol:** Dosing modeled after Herceptin (trastuzumab), an established HER2-targeted biologic: 4 mg/kg loading dose followed by 2 mg/kg weekly maintenance doses (Roche 2025). Assume 12 month-treatment cycle and 70 kg average patient weight, yielding 7,420 mg total drug exposure per patient (280 mg loading dose + 7,140 mg maintenance doses over 52 weeks)

- ⌃ **Pricing/Revenue Expectations:** $4.66 per mg based on Herceptin's 2005 AWP ($2,930 for 440mg (Gemmete and Mukherji 2011; Fleming 2005), adjusted to $2,051 ASP for 440mg), equating to $34,587 per patient in 2005 USD

- ⌃ **SAM Calculation:** 33,929 treated patients × $34,587 revenue per patient = $1.2B in 2005 USD ($1.9B in 2025 USD)

Valuation Insight: The bottom-up analysis revealed a SAM of approximately $1.2B in 2005 USD ($1.9B in 2025 USD). This market sizing approach helped Company A understand how patient population filters (disease subtype, staging, market share, coverage, and adherence) significantly impact addressable market size, dropping from 4 million total breast cancer patients to 33,929 treated patients—less than 1% of the total population.

Why It Matters: This case demonstrates how biotech companies must layer multiple real-world constraints when sizing specialized oncology markets. The analysis shows that even with a large disease population (4 million), targeting a specific biomarker (HER2-positive) and accounting for practical factors like insurance coverage and patient adherence dramatically reduces the addressable market. A transparent methodology is critical for accurate valuation, as small adjustments to variable inputs can significantly impact revenue projections. This underscores the importance of using current industry benchmarks and reliable comparable data to inform key development priorities and commercial strategy decisions.

Calculation Limitations and Considerations

Of course, this is a rough calculation for illustrative purposes. Here are the key assumptions and simplifications made in this example:

1. **Eligibility and Adherence**: We based our calculations on the assumption that patients with HER2-positive, non-metastatic breast cancer have a 90% treatment rate, 60% coverage rate, and 85% adherence rate. In reality, additional factors will influence eligibility such as treatment history, comorbidities, etc. In addition, not all eligible patients will necessarily receive the therapy due to factors such as therapy preferences, physician recommendations, and patient preferences. Furthermore, adherence can be affected by various factors, including side effects, treatment convenience (oral versus IV infusion), and disease severity.

2. **Market Dynamics**: The 20% market share estimate is purely hypothetical and depends on many variables, including competition, market entry barriers, order of entry, and shifts in treatment guidelines. Changes in the competitive landscape and the introduction of new therapies could further impact this share.

3. **Pricing and Reimbursement:** The revenue calculation is based on historical pricing of Herceptin in 2005, which may not be indicative of future pricing trends. Current pricing strategies, discounts, and reimbursement policies need to be considered when calculating revenue.

4. **Treatment Duration and Efficacy:** The revenue estimate assumes that all patients will receive the biologic for 12 months. However, actual treatment duration may vary based on clinical outcomes, disease progression, and individual patient needs. Consulting a medical professional or oncologist for the latest practices would be helpful.

5. **Regulatory and Market Access:** Regulatory approvals, market access challenges, and health policy changes can affect the availability and uptake of the biologic, which we did not consider in this example.

What to Keep in Mind

• • • • • • • • • •

Being first to market gives you an edge (about 6% more market share than you'd otherwise expect), but it's not the whole story. As we've seen with breakthrough treatments like Tagrisso and Eliquis, having a truly differentiated product can help you leapfrog competitors even when you're fashionably late to the party.

Take a moment to honestly reflect on your product's unique value. What will make physicians choose it? Will patients notice meaningful improvements in their lives? The market share you ultimately capture isn't set in stone—it evolves with the strategic choices you make, from how you design clinical trials to how you tell your product's story after launch.

Understanding these market dynamics gives you practical leverage to maximize your product's potential, even when you're not the pioneer. As we move into the next chapter, we'll get even more specific about the biotech journey, diving into launch curves, exclusivity timelines, commercialization approaches, and manufacturing considerations that shape your financial projections.

BEFORE YOU WRAP UP

Take a moment to reflect on what you've learned. These key takeaways and questions are designed to help you connect the concepts to your own goals and identify areas you'd like to explore further.

Key Takeaways

- Market opportunity assessment is not just about market size—it's about understanding how your innovation addresses unmet needs, differentiates from competitors, and can sustainably grow in the market.

- Total Addressable Patient Population (TAPP) forms the foundation of market sizing and requires careful consideration of disease prevalence/incidence, diagnosis rates, treatment rates, and patient adherence/compliance.

- Market share is significantly influenced by both order of entry and product differentiation—first-in-class products typically retain larger market share, but truly differentiated "best-in-class" products can overcome the timing disadvantage.

- Drug pricing is complex, with critical differences between list prices (wholesale acquisition cost or WAC) and what manufacturers actually receive (net price, approximated by average sales price or ASP).

- International markets typically price drugs at only about 36% of U.S. prices, requiring significant adjustments in revenue projections, depending on the region.

Reflection Questions

1. Which aspect of sizing the TAPP do you think is most commonly overlooked or underestimated by early-stage biotech companies?

2. How might your valuation approach change if your product will be third-to-market but offers significant advantages in either safety or route of administration?

3. What surprised you most about drug pricing dynamics, particularly regarding the gross-to-net calculations and their impact on manufacturer revenue?

4. How would you approach modeling patient adherence for a novel therapy in your specific therapeutic area?

5. How could understanding the international pricing landscape influence your early development and commercialization strategy?

6. Which market opportunity factor do you believe will be most challenging to estimate for your specific product, and what resources might help you address this challenge?

03

Product Dynamics

E ven if a therapy has a large addressable market and cutting-edge technological platform, its ultimate success hinges on commercial performance. A strong drug launch is essential for creating value and, more importantly, sustainability and long-term growth of biotech companies, especially given the substantial R&D investments made with the expectation that market success will drive profitability. This principle applies not only in biotech but across all business sectors.

In this chapter, we'll delve into some fundamental concepts of product dynamics, including drug launch curves, LOE, commercialization, and Cost of Goods Sold (COGS). These elements will form the backbone of our biotech valuation model and highlight the vital role of successful drug launches as revenue drivers for an early-stage biotech company.

Launching a New Drug to Market

The launch of a novel drug or biotech product comes with the advantage of market exclusivity, granting the manufacturer the sole right to sell the product and blocking generic competitors from entering the market. This advantage stems from both regulatory protections tied to the initial application process and patent safeguards that protect the underlying intellectual property (IP). This exclusivity period is critical, allowing the company to recoup the substantial costs incurred during R&D,

clinical trials, and regulatory approvals. However, bringing a new therapeutic to market is never without challenges. Generic manufacturers frequently pursue legal strategies to bypass exclusivity or patent protections, often leading to litigation over patent infringement or attempts to bypass exclusivity laws. Moreover, even before these legal battles arise, getting new therapeutics into the hands of providers and patients has become increasingly difficult in the present days.

An analysis of drug launches into the market from 2006 to 2013 showed that the average performance of launches has declined for three "success" metrics which included peak revenues, the shape of the revenue curve, and the percentage of analysts' launch expectations achieved after 5 years (Donoghoe et al. 2017). Multiple conclusions can be drawn from this to why launches have been more difficult in modern times: increasing competition, reduced chance of achieving a significant increase in clinical benefit, payers restricting market access, and public awareness of drug price hikes by pharma companies. Many factors influence a successful drug launch, and it may be helpful to consider four common launch categories, which will help guide the scenarios to consider in your asset valuation:

01 Breakthrough Drugs

Rare but transformative treatments that revolutionize care and are likely to sustain strong market success.

02 Drugs with Significant Benefits and Some Drawbacks

These therapies may have limitations—such as safety concerns or inconvenient delivery—but can still thrive if companies effectively demonstrate their value.

03 Drugs with Limited Clinical Benefits but Other Clear Advantages

These treatments may not significantly outperform the standard of care clinically but can succeed through advantages like lower cost, improved accessibility, or regulatory pathways (e.g., generics, biosimilars, or 505(b)(2) approvals).

04 Drugs Unlikely to Compete

Treatments with minimal clinical or market advantages should be identified early and deprioritized to avoid costly late-stage development. If a drug is unlikely to compete due to a lack of meaningful benefits, whether clinical, cost-related, or regulatory, it may be more strategic to halt further investment.

As discussed in Chapter 2, being first to market offers a competitive edge, and companies often invest heavily to improve their chances of beating competitors to launch. Of course, being first does not necessarily guarantee success as there are other factors to consider such as best-in-class, number of competitors, therapeutic benefit, etc. Moreover, late entrants are not doomed to lose market share, as 40% of market leaders were found to be late entrants (Cha and Yu 2014). These companies can boost their chances of success by spending resources to keep pace with early leaders and devising key strategies to differentiate their products (Allmendinger 2021).

With these points in mind, it's crucial to understand how investors, dealmakers, and executives are seeking the next big investment opportunity. That's why you must present a compelling, data-driven valuation model that convincingly showcases why your venture is worth their investment, which is ultimately tied to strong market performance.

To evaluate market performance of a new drug, start by answering two critical questions:

1. How much revenue will it generate over time?
2. How long will it take to reach peak sales or revenue?

Our valuation model is designed to address these very questions by using what the industry calls "launch curves," a tool that allows you to predict how a drug will perform once it hits the market.

Understanding Launch Curves

Launch curves offer valuable insight into how drugs generally behave after entering the market and have continuously updated by drug launch modeling experts (Bauer and Fischer 2000; Robey and David 2017; Teramae et al. 2020). These models allow us to predict sales growth and the time of peak revenue, even when a drug is still years away from clinical trials or market approval.

Some investors and executives prefer to tailor their launch curves to reflect the unique characteristics of a specific drug or market, rather than use a standardized model. However, research into these factors is often limited, as available data can quickly become outdated due to how rapidly the drug industry changes, including shifts in global environment, competition, and regulatory space.

To address these challenges, we present a robust base case valuation model that allows for simple-to-understand customization based on your specific drug or market factors. This flexible framework ensures that our model can be applied to a diverse array of drugs, enabling entrepreneurs, investors, and executives to quickly obtain valuable insights without getting bogged down by the endless market and research data. This is especially true for early-stage companies where R&D can extend over many years and it would be more practical to focus on general characteristics to develop a solid foundation rather than striving for a perfect estimate.

So, how do we go about creating a launch curve?

The launch curve of a drug offers both a visual and quantifiable representation of how sales typically increase each year from the designated launch date until they reach peak annual revenue, which is directly influenced by TAPP and market share (Robey and David 2017), as discussed in Chapter 2.

When constructing a launch curve, we must consider two essential parameters:

1. **Time to Peak:** The time it takes (duration) for a drug to achieve peak market penetration or maximum revenue.
   - In markets with low competition and strong marketing support, a product may achieve maximum revenue more quickly.

2. **Shape of the Curve:** The steepness and trajectory of the uptake curve.
   - If a market adopts a drug rapidly, we can anticipate a steeper uptake curve and a trajectory that suggests higher revenue.

Now, let's take a closer look at each of these key parameters to consider when building your own launch curve.

Time to Peak

In 2000, Bauer and Fischer proposed that early innovators or pioneers tend to reach peak sales more slowly as adoption of novel therapies requires convincing practitioners and overcoming resistance to change (Bauer and Fischer 2000). This observation can vary between physician groups and specialties, reflecting differences in clinical practice and therapeutic needs. For instance, oncologists are more likely to adopt new therapeutic products due to urgent clinical needs. Figure 17 demonstrates the general sales trajectories of pioneer/early mover drugs and late movers with time to peak at 9 years and 3 years post-launch, respectively (Bauer and Fischer 2000).

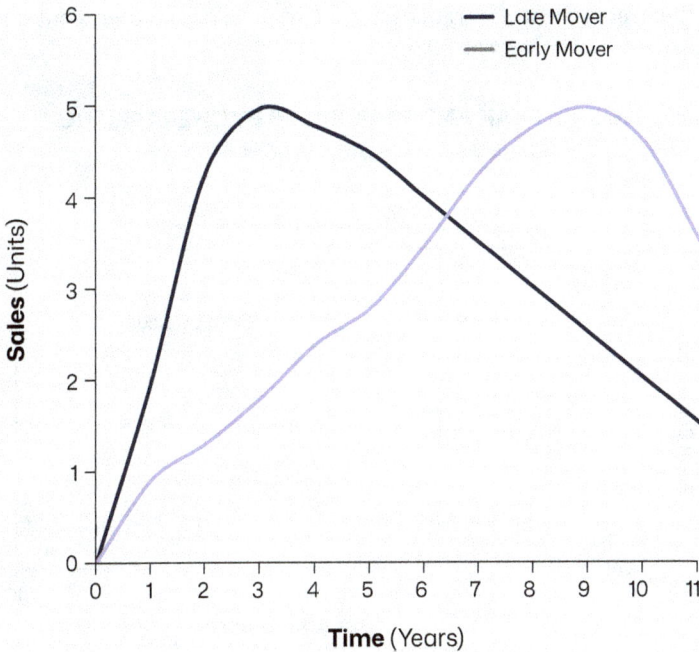

LAUNCH CURVE COMPARISON: EARLY MOVER VS. LATE MOVER PERFORMANCE

Adapted by AGMI Group from Bauer and Fischer 2000

Figure 17. Comparison of typical market launch curves for early mover versus late mover products.

Recently, further refinement of time-to-peak estimates has become an important focus, as both the timing of peak sales and the subsequent decline can significantly impact overall revenue potential. A 2020 study analyzed the sales trajectories of 272 top-selling drugs (those with annual global sales exceeding $2B) between 2011 and 2017 (Teramae et al. 2020). The findings revealed a median time to peak sales of approximately eight years, with a range of 4 to 12 years.

Further analysis of drug life cycles revealed notable variations across therapeutic areas and modalities. Infectious disease drugs reached peak sales the fastest, averaging around four years, while endocrine and metabolic drugs peaked much later, at approximately 11 years. Oncology drugs fell in between, reaching their peak in about 7 years. A similar pattern emerged when comparing drug modalities: small-molecule drugs typically peaked within 8 years, whereas biologics took closer to 11 years to peak.

Figure 18 illustrates these differences in time to peak across therapeutic areas and drug modalities (Teramae et al. 2020).

TIME TO PEAK AND PEAK SALES BY THERAPEUTIC AREA AND MODALITY, 2011 TO 2017

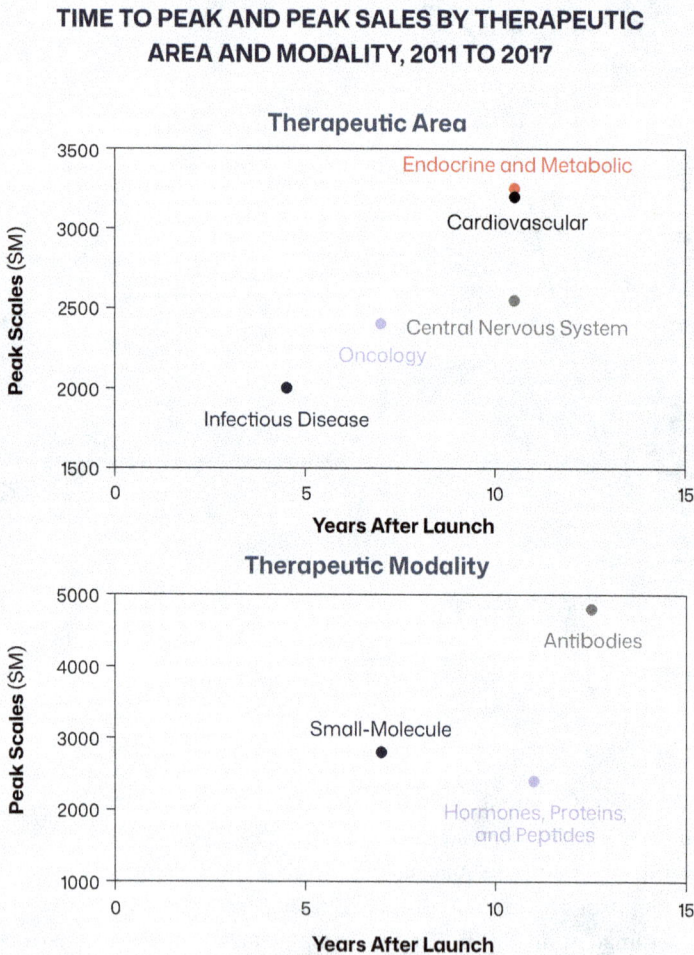

Source: Teramae et al. 2020

Figure 18. Time to peak performance by therapeutic area and modality of top-selling drugs from 2011 to 2017.

In terms of market entry, first-mover and early-mover drugs often enjoy a competitive advantage, but this trend is not consistently observed across all cases. Currently, there is no clear pattern linking this advantage to specific therapeutic areas or modalities. A 2017 study published in Nature Discovery supports this, showing no significant differences in sales trajectories between pioneers and followers, which contrasts with Bauer and Fischer's findings (Robey and David 2017). Furthermore, they did not observe a difference between biologics and non-biologics. These findings suggest that pioneer and early-mover drugs do not consistently differ from their later-entering competitors.

Several factors could explain this:

- **Enhanced Marketing Reach:** Modern marketing techniques have made widespread dissemination of information more accessible, allowing companies to conduct targeted awareness activities and market education across specific physician segments. This enables companies to educate and influence provider and patient behavior as well as drive prescriptions for newer therapeutics (Morrison 2020).

- **Increased Competition:** Companies now face more intense competition with less differentiated products in crowded therapeutic areas related to increasing difficulty in scientific and medical breakthroughs (Mandell and Hattem 2019).

- **Payer Pressure:** Negotiations with payers have become increasingly challenging as they continue to raise their standards for coverage and reduce reimbursement levels (Pearson et al. 2020).

Based on the latest analysis, we use a base case of 8 years for most early-stage drugs to reach their peak sales or revenue. We can adjust this depending on other factors that we will discuss a little later, which can then be applied to a diverse set of drugs to reflect the breadth of various specific drug characteristics, therapeutic markets, and drug launch scenarios.

As an aside, the post-COVID-19 era has highlighted how urgent clinical needs at global scale can accelerate the development and adoption of effective treatments in specific scenarios. The pandemic spurred the rapid creation and global distribution of mRNA vaccines, which represented a revolutionary alternative to traditional vaccine strategies. These vaccines stood out for their high potency, scalability, and potential for cost-effective production and safe administration (Mittal 2024).

Typically, it takes years to move from platform development to therapeutic application, but COVID-19 demonstrated how a novel treatment could be rapidly adopted while maintaining scientific rigor and patient safety. This has revitalized efforts to strengthen collaboration, expand digital health platforms, and encourage regulatory bodies to adapt their processes to support the swift development and approval of safe and effective therapeutics (Saikia et al. 2023).

In addition, the pandemic has accelerated the use of innovative technologies such as antibody engineering, RNA interference, AI, and adaptive trial designs. As of this guide's publication, it remains unclear whether these advancements will affect the time to peak adoption of new treatments. It's possible that certain conditions must be met for such rapid and widespread adoption, including a significant clinical need, the severity of the disease, number of impacted individuals, and the nature of the breakthrough technology. Only time will tell.

> **Quick tip:** The model is particularly sensitive to the time to peak revenue, as it significantly impacts the "time value of money" and, therefore, your valuation potential. We will explore this topic in greater detail when we start building our SOTP valuation model and adjusting variables in Chapter 6.

Variable 2

Shape of the Curve

The shape of a launch curve follows three different phases of product diffusion:

1. **Slow Initial Uptake from Early Adopters:** Early adopters are driven by factors such as limited treatment options, novel mechanisms of action, or unmet medical needs. In addition, market education and initial targeting of key opinion leaders is important for gaining the influence required to change the practice of medicine (e.g., general practitioners take the lead from specialists, specialists take the lead from KOLs).

2. **Rapid Diffusion Among the Majority:** Market penetration accelerates as real-world patient outcomes are observed, reimbursement approvals are secured, and outreach efforts by the company—along with peer recommendations among physicians—gain traction.

3. **Tapering Off with Late Adopters:** By this stage, the drug is typically well-established with a solid understanding of its long-term safety and efficacy. Late adopters are often influenced by factors such as cost-effectiveness, insurance reimbursement, and extensive post-commercialization clinical data. Biotech companies shift their focus to sustaining market share, protecting against generic competition, and managing lifecycle strategies like new indications or formulations.

Based on these phases, the launch curve forms an S-shape, which was first modeled and described by Bauer and Fischer (2000) for pharma drugs and further reaffirmed by more modern drug uptake industry analyses. The shape can be further modified by making adjustments by left- and right- shifting the curve, which will accelerate or delay, respectively, the uptake of the drug in the market.

Unifying Launch Curve Variables

By combining the two key variables—time to peak and the curve shape—we create a product launch curve that can be adapted to most biotech valuation analyses. Figure 19 illustrates an S-shaped launch curve with an 8-year time to peak, showing how sales evolve as a percentage of the peak over the years following product launch, based on analysis by Robey and David (2017) and Teramae et al. (2020). The median values and interquartile ranges of the percent peak sales enable adjustments to the curve's shape. For example, if a slower uptake is anticipated, the 75th percentile curve may be a more appropriate fit. On the other hand, if rapid uptake is expected, the 25th percentile curve might better reflect the trajectory.

TYPICAL DRUG LAUNCH CURVE

Years After Launch	Median % of Peak Sales	Interquartile Range
1	8	3-13
2	21	12-30
3	37	24-48
4	58	41-67
5	71	60-81
6	83	77-91
7	92	88-96
8	100	n/a

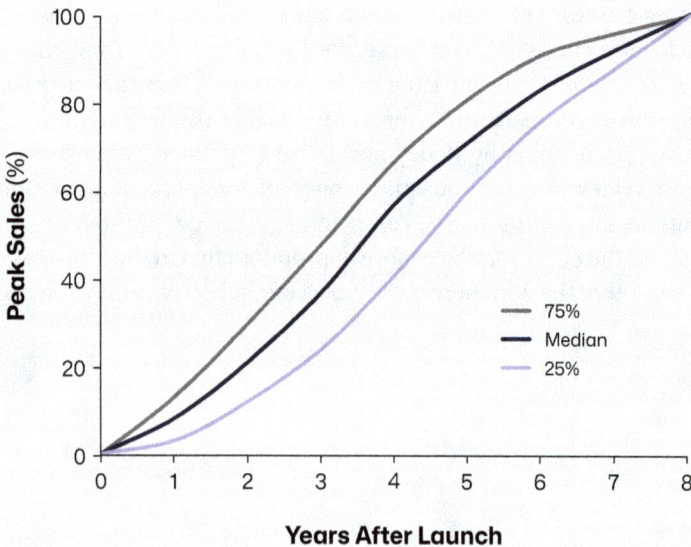

Adapted by AGMI Group from Robey and David 2017 and Teramae et al. 2020

Figure 19. *Typical drug launch curve showing 8-year time to peak, with median and 25th/75th percentile curves. Included table shows median and interquartile range by year.*

Customizing your Launch Curve Variables

Adjusting key variables like how soon a new therapeutic reaches peak sales (time to peak) and how quickly sales ramp up (the shape of the launch curve) directly impacts your biotech product's valuation. Since these factors can either increase or decrease projected value, it's essential to base your estimates on well-supported assumptions backed by reliable industry benchmarks or thorough analysis. The perceived level of demand plays a crucial role in shaping your launch curve, and several key factors should be considered when making these adjustments.

Accelerate or Delay the Uptake Curve

- **Pricing and Reimbursement Strategy**: A strong pricing strategy and favorable reimbursement status can drive faster uptake by providers, mitigating temporary market access barriers.

  - Note that reimbursement delays often slow adoption rates rather than reduce overall product opportunity.

- **Unmet Clinical Need and Market Receptiveness:** When a drug addresses significant unmet clinical needs (i.e., addresses the needs of the underlying disease and patient population), uptake accelerates dramatically.

  - Different physician specialists are more receptive to the latest therapeutics such as oncologists due to very high clinical needs.

  - For example, imatinib extended life expectancy from chronic myeloid leukemia from 2 to 3 years to over 10 years.

- **Therapeutic Benefit**: An innovative drug with superior clinical effectiveness is likely to be adopted more quickly by physicians and patients.

- **Product Differentiation**: Lack of differentiation such as efficacy, safety, convenience, or indication may delay adoption as physicians stick to trusted alternatives.

- **Competition**: The presence of strong or established competitors, brand recognition, or lower pricing can slow the uptake of a new drug. Without a competitive edge, the product may struggle to gain early traction in the market.

⊿ **Healthcare Cost Savings:** A new drug that carries a higher price point but reduces overall healthcare costs may accelerate adoption.

- ◆ For example, a therapy that significantly reduces hospital stays, emergency visits, or long-term disease complications can be attractive to payers and providers despite its upfront cost. Additionally, drugs that minimize "leakage costs," such as reducing the need for additional interventions or hospital readmissions, can drive faster market uptake by demonstrating clear economic benefits.

Increase or Decrease the Time to Peak

⊿ **Follow Similar Launch Patterns**: By modeling uptake patterns after similar drugs with known market trajectories, one can estimate the time to peak. Drugs that resemble existing therapies or enter saturated markets may experience a slower rise to peak.

⊿ **Model a Family of Products**: Expansions of the product line or broader indications (i.e., label expansions) may extend the time to peak, offering a longer growth period and delaying peak sales.

⊿ **Early Versus Late Market Entry**: While we typically don't differentiate between pioneers/early adopters and followers, pioneer drugs in certain markets may experience a longer time to peak as physicians are more hesitant to adopt and trust in a new therapeutic approach, especially therapies with a new technological platform. In contrast, follower drugs often reach their peak faster by leveraging the established trust and familiarity of earlier, similar treatments.

⊿ **Biologics Versus Non-Biologics:** Biologics may have a longer time to peak compared to non-biologics, although this remains a topic of debate (Robey and David 2017; Teramae et al. 2020).

- ◆ For example, biologics often involve complex manufacturing, specialized administration (e.g., injections or infusions), and higher costs, all of which could slow adoption and eventual time to peak. Additionally, physicians and payers may take longer to adopt biologics due to concerns about efficacy compared to non-biologic alternatives, drug costs, reimbursement challenges, and the need for prior authorizations.

- ◆ In contrast, non-biologics, especially oral small molecules, often may have a shorter time to peak due to easier prescribing, method of administration, broader accessibility, and lower cost barriers.

Adjusting these two variables can be time consuming when modeling launch curves from scratch. To facilitate analysis, Table 5 synthesizes data from multiple studies, enabling rapid assessment of various curve types under different time-to-peak scenarios and uptake speeds through median and interquartile ranges of peak sales percentages (Robey and David 2017; Teramae et al. 2020).

DRUG LAUNCH CURVES WITH VARYING TIME TO PEAK

Median (interquartile range)

Time to Peak (Year)	Uptake Length (Years)								
	4	**5**	**6**	**7**	**8**	**9**	**10**	**11**	**12**
Year #1	21% (14-30%)	15% (7-25%)	11% (5-21%)	9% (4-18%)	8% (3-13%)	7% (3-11%)	6% (2-11%)	5% (2-9%)	5% (2-8%)
Year #2	58% (41-67%)	42% (27-53%)	31% (20-41%)	25% (16-35%)	21% (12-30%)	18% (11-26%)	15% (7-24%)	13% (6-22%)	12% (5-20%)
Year #3	83% (78-91%)	68% (57-78%)	58% (41-67%)	46% (30-57%)	37% (24-48%)	31% (20-41%)	28% (18-37%)	25% (15-34%)	22% (13-32%)
Year #4	100%	86% (82-94%)	76% (66-85%)	66% (52-75%)	58% (41-67%)	49% (33-59%)	42% (29-53%)	38% (27-48%)	35% (24-45%)
Year #5		100%	89% (85-96%)	80% (73-88%)	71% (60-81%)	64% (49-73%)	58% (41-67%)	50% (38-62%)	47% (35-58%)
Year #6			100%	90% (87-96%)	83% (77-91%)	76% (66-85%)	69% (57-78%)	62% (50-74%)	58% (47-69%)
Year #7				100%	92% (88-96%)	85% (79-93%)	79% (72-87%)	72% (64-82%)	68% (60-77%)
Year #8					100%	93% (89-97%)	86% (82-94%)	80% (75-88%)	76% (70-85%)
Year #9						100%	93% (90-97%)	87% (83-93%)	83% (79-90%)
Year #10							100%	93% (90-97%)	90% (86-95%)
Year #11								100%	95% (92-98%)
Year #12									100%

Note: These are example drug launch curves provided for reference. The length and shape of a launch curve can vary significantly depending on factors such as the therapeutic area and market environment. Uptake length 11 and 12 are extrapolated based on data from uptake length 4 to 10.

Adapted by AGMI Group from Robey and David 2017 and Teramae et al. 2020

Table 5. Drug launch curves showing varying time to peak, with median and 25th/75th percentile ranges.

Effects of Loss of Exclusivity (LOE)

● ● ● ● ● ● ● ● ● ●

When a biotech product's patent protection expires, generic or biosimilar competitors can enter the market, resulting in a sharp decline in the drug's pricing and sales during a critical period known as LOE. This transition is facilitated by U.S. federal laws, such as the Hatch-Waxman Act and the Biologics Price Competition and Innovation Act (BPCIA), which establish pathways for generic and biosimilar manufacturers of small-molecule drugs and biologics, respectively, to apply for FDA approval for their products, even before the expiration of the brand-name drug's patents (Grabowski et al. 2021; Aitken, Kleinrock, and Muñoz 2020). Figure 20 shows the timed entry of generics entering the market under the Hatch-Waxman Act (PhRMA 2024).

HATCH-WAXMAN TIMELINE OF SMALL-MOLECULE DRUG COMPETITION

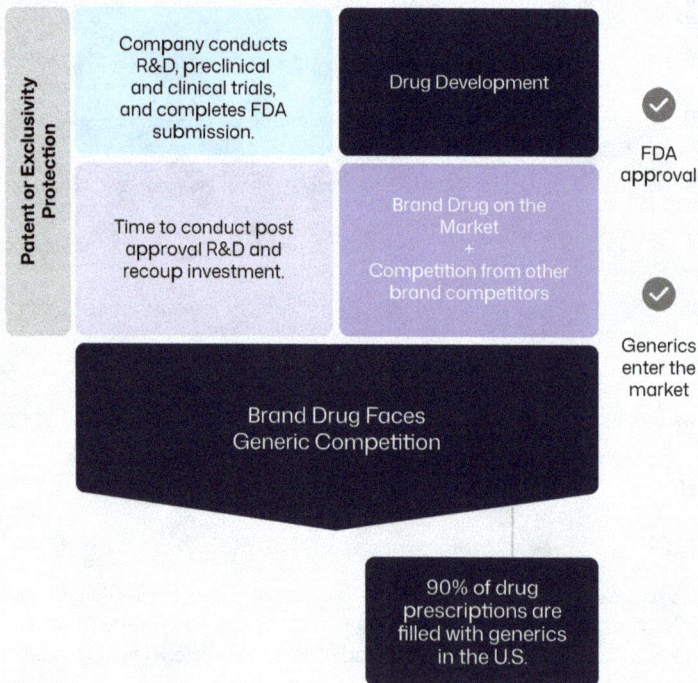

Patent or Exclusivity Protection

Company conducts R&D, preclinical and clinical trials, and completes FDA submission.

Drug Development

✓ FDA approval

Time to conduct post approval R&D and recoup investment.

Brand Drug on the Market
+
Competition from other brand competitors

✓ Generics enter the market

Brand Drug Faces Generic Competition

90% of drug prescriptions are filled with generics in the U.S.

Adapted by AGMI Group from PhRMA 2024

Figure 20. Timeline illustrating the transition from brand-name to generic drug competition for small-molecule therapies under the Hatch-Waxman Act.

These laws have also streamlined the approval process by allowing generic companies to reference the safety and efficacy data of brand-name products, reducing development costs and time to market. Consequently, these companies can prepare to launch their products through successful patent challenges, settlement agreements allowing earlier entry, or as soon as patents and exclusivity protection expire, further accelerating competition and price erosion in the market. Overall, these laws have played a pivotal role in expanding the generics and biosimilars market with the primary purpose of improving access to care and significantly lowering costs nationwide, all while maintaining the incentives to motivate continued innovation of new therapeutics in the biotech and pharma industry. Annual savings from generics and biosimilars exceeded $445B in 2023, more than $12B of which comes from biosimilars (Association for Accessible Medicines 2024). Because of their lower costs, generics and biosimilars represent more than 90% of the prescriptions dispensed in the U.S. Figure 21 provides an overview of the increased savings associated with the generic and biosimilar manufacturers.

ANNUAL SAVINGS FROM GENERICS AND BIOSIMILARS BY MARKET ENTRY PERIOD, 2014 TO 2023

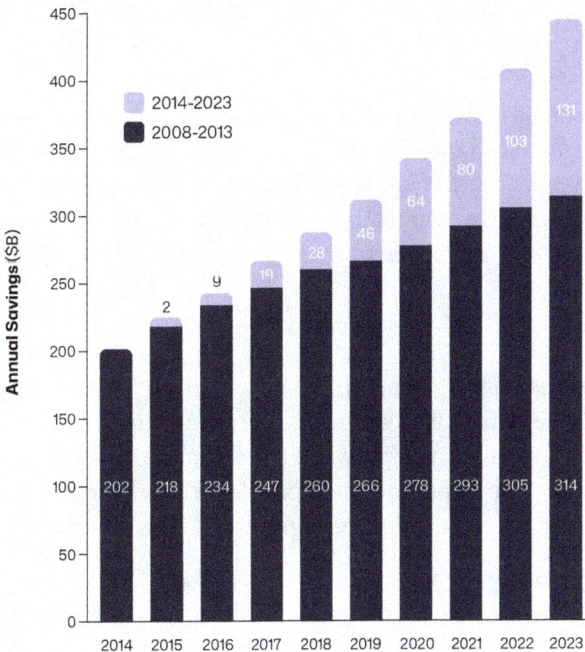

Source: Association for Accessible Medicines 2024

Figure 21. Annual healthcare savings from generic and biosimilar drugs from 2014 to 2023 segmented by their U.S. market entry period.

Although LOE is typically far in the future and has minimal effect on current valuations, biotech models are still expected to include an analysis of the significant revenue decline that follows LOE, when generics enter the market and fierce competition begins. There are three factors to consider when modeling LOE: timing of competition, market share after LOE, and price after LOE.

Timing of Competition

Once the patents or exclusivity protection that safeguard a drug's market exclusivity expire, the drug's revenue potential declines significantly due to the entry of generic or biosimilar competitors. A recent analysis of market exclusivity periods for new drugs between 2009 and 2019 found that both small-molecule and biologic drugs enjoyed protection ranging from 10 to 17 years, with an average of 14 years (Grabowski et al. 2021). Biologics typically enjoyed extended protection, averaging 7 additional years compared to small molecules (Rome et al. 2021). These findings align with earlier large-scale analyses (Wang et al. 2015). When specific patent information is unknown or unavailable for a drug candidate, we generally estimate approximately 14 years of market exclusivity from launch until competition from generics or biosimilars.

It's possible to estimate when a drug might face competition based on its patent term and government-granted market protection, but doing so is difficult without detailed information from the patent-holding company or experts in pharmaceutical intellectual property law. Estimating the period of market exclusivity requires consideration of the following factors:

PATENT TERMS

Patents typically last for 20 years in the U.S. starting from the application filing date, but effective patent life can be shorter due to the time taken for clinical development and regulatory approval (U.S. Food and Drug Administration 2020). Note that drugs have multiple patents in parallel to cover various aspects including composition, formulation, manufacturing, method of use, etc.

- Patent information for drugs in development is available through public databases such as the U.S. Patent and Trademark Office (2025) or in the financial reports of publicly traded companies. However, do note that figuring out which patents actually control market exclusivity usually involves insights from experts. For drugs already on the market, you can easily check this in the FDA Orange Book U.S. Food and Drug Administration (2025).

MARKET EXCLUSIVITY

FDA-granted market protection periods that prevent approval of competing generic or biosimilar products for specified timeframes, operating independently of patent rights.

- **Small-Molecule Drugs (under the Hatch-Waxman Act):** Regulatory exclusivity framework to encourage small-molecule innovation with timely generic drug market entry.

  - **New Chemical Entity (NCE) Exclusivity:** 5-year exclusivity period from the date of approval for the brand-name small-molecule drug with a new active moiety from the entrance of generics (PhRMA 2024). This exclusivity is independent of patents and starts at approval, not patent filing.

    After 4 years, a generic application can be submitted only with a Paragraph IV patent challenge (i.e., generic manufacturer states that patent is invalid, unenforceable, or won't be infringed by the generic).

  - **New Clinical Investigation (NCI) Exclusivity:** 3-year exclusivity period for certain changes to already-approved active moiety—such as new indications, formulations, or dosing regimens—when the approval is supported by new clinical studies conducted or sponsored by the applicant that are essential to the change (PhRMA 2024). This exclusivity runs concurrently with any patent terms and does not block generics of the original product, only of the newly approved change.

- **Biological Products (under the BPCIA):** Regulatory exclusivity framework to encourage biological product innovation while eventually allowing biosimilar competition.

  - **Biologic Exclusivity:** 12-year exclusivity period from the date of approval for the brand-name biological product from the entrance of biosimilars (Bloomberg Law 2011). This exclusivity is independent of patents and starts at approval, not patent filing.

    After 4 years, a biosimilar application can be submitted, but approval is barred until 12 years have passed.

⋏ **Additional Exclusivity Programs** (can be submitted to the FDA to modify or extend exclusivity periods): Supplementary incentive programs designed to encourage development in underserved therapeutic areas and patient populations (Patel 2023).

- **Orphan Drug Exclusivity**: 7-year exclusivity period for drugs that treat specific rare diseases or conditions affecting fewer than 200,000 people in the U.S. This exclusivity applies to both small molecules and biologics and runs independently of patents or another exclusivity.

- **Pediatric Exclusivity**: Additional 6 months of exclusivity if the sponsor conducts FDA-requested pediatric studies. This adds on top of any existing exclusivity or patent term, for both small molecules and biologics.

- **Antibiotic Exclusivity**: Additional 5 years of exclusivity for drugs designated as Qualified Infectious Disease Products (QIDP). This is for the purposes of incentivizing the development of antibacterial and antifungal drugs for human use to treat serious or life-threatening infections. This exclusivity is added on top of any existing exclusivity periods or patent protections. Note that this is only applicable to small-molecule drugs, not biologics.

⋏ **Strategic Exclusivity Extensions**: Companies may extend market exclusivity by pursuing strategies such as developing new formulations, indications, methods of use, or combination therapies—approaches that apply to both small molecules and biologics, though the regulatory pathways and resulting exclusivity periods differ between the two.

Figure 22 summarizes the exclusivity periods for each different FDA application (U.S. Food and Drug Administration 2020).

OVERVIEW OF COMMON DRUG EXCLUSIVITY PERIODS

SM = Small Molecule

B = Biological Product

■ **5 Years**
SM

New Chemical Entity (NCE) Exclusivity

A brand-name drug with a new active moiety has a **5-year** exclusivity

■ **3 Years**
SM

New Clinical Investigation (NCI) Exclusivity

A brand-name small-molecule drug with a previously approved active moiety has a **3-year** exclusivity for certain changes, such as new indications, formulations, or dosing regimens, when the approval is supported by additional clinical studies.

■ **7 Years**
SM   B

Orphan Drug Exclusivity (ODE)

A brand-name drug for a disease or condition that affects fewer than 200,000 people in the U.S. (or that affects more people but for which the drug company still has no hope of covering the development costs) has a **7-year** exclusivity.

■ **12 Years**
B

Biologic Exclusivity*

A brand-name biological product, including traditional biologics and cell and gene therapies, has a **12-year** exclusivity.

Add-On Programs

+ 6 months
SM   B

Pediatric: A brand-name drug for which the sponsor has done pediatric studies (in response to a written request from FDA) may be eligible for a 6-month exclusivity, which is added on to any other exclusivities or patents for that drug..

+ 5 years
SM

Antibiotic: Certain new antibiotic drugs for specific infectious diseases may be eligible for a 5-year exclusivity, which is added on to any other exclusivities or patents for that drug.

Biologics as defined by FDA include cell and gene therapies, RNA-based therapies, as well as proteins, monoclonal antibodies, and vaccines that are derived from living organisms and require complex manufacturing processes.

Adapted by AGMI Group from U.S. Food and Drug Administration 2020

Figure 22. Overview of standard exclusivity period granted under FDA regulations for therapeutic market protection.

> **Quick tip:** If you believe a major competitor will significantly erode market share despite your product having market exclusivity, consider modeling their entry as if it occurs at LOE rather than waiting for the actual patent expiration or generic/biosimilar entry.

Impacts on Revenue

There will be substantial revenue decline for a drug once patents expire and/or any exclusivity period ends. As previously stated, the entrance of drug generics and biosimilars into the market will lead to increased competition and cause a decline in both pricing and volume. Figure 23 demonstrates the total revenue loss for both brand-name biologics and small molecules from LOE in the U.S. from 2019 to 2023 (forecasted to 2028) (IQVIA 2024). Biologics are expected to experience about $40B of brand losses over from 2022 to 2026 related to the LOE and introduction of biosimilars for brands such as Lucentis (ranibizumab) from 2022, Humira (adalimumab) from 2023, and Stelara (ustekinumab) from 2025 (Rudge 2022; Merrill 2022)..

LOSS OF EXCLUSIVITY IMPACT ON U.S. DRUG BRANDS, 2019 TO 2028

Total Brand Loss due to LOE ($B)

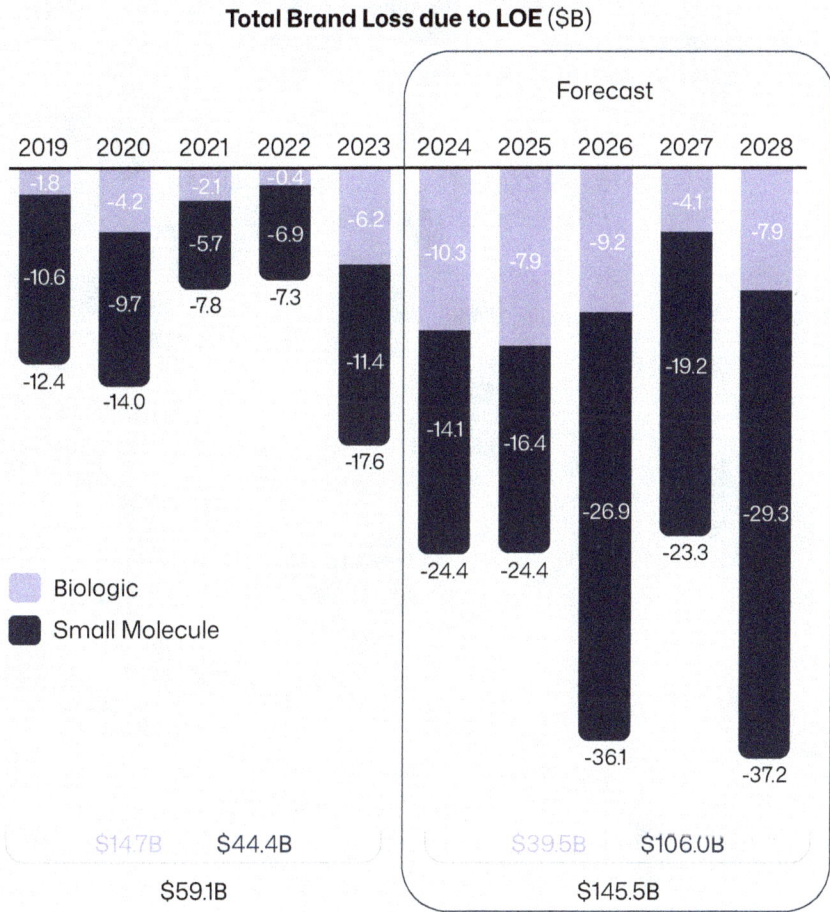

Source: IQVIA 2024

Figure 23. Impact of LOE on brand drugs between small molecules and biologics from 2019 to 2028.

Since the early 2000s, generics have claimed a significant share of the market, rapidly displacing brand-name counterparts. For new molecular entities (NMEs)—products with active ingredients marketed for the first time in the U.S.—that faced generic competition between 2017 and 2019, brand-name products retained only 23% of the market after one year (Grabowski et al. 2021). The impact was even more pronounced for blockbuster drugs (those generating over $250M in 2008 dollars), where brand-name retention dropped to just 18%. In contrast, back in 1999 to 2000, brand-name drugs still held 44% of the market one year after generics entered. This stark decline underscores the accelerating adoption of generics over the past two decades, as seen in Figure 24 (Grabowski et al. 2021).

BRAND MARKET SHARE RETENTION AFTER GENERIC ENTRY, 1999 TO 2019

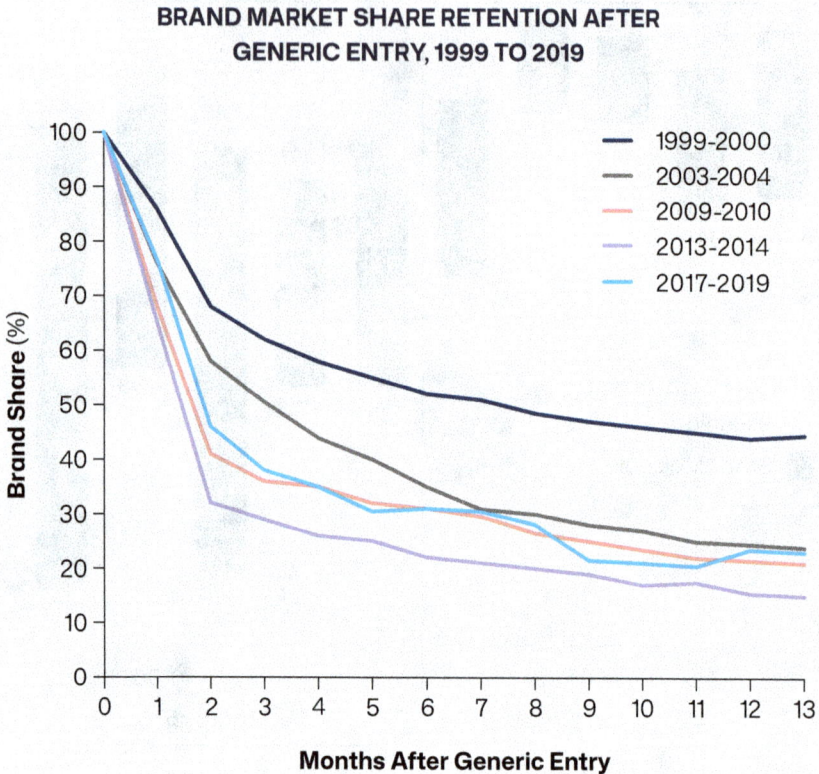

Note: Data includes New Molecular Entities (NMEs) only. Select years from 1999 to 2019 were chosen for visual purposes. Brand share is calculated as a product's sales volume divided by the total market sales volume, expressed as a percentage.

Source: Grabowski et al. 2021

Figure 24. Monthly brand share performance for small molecules following first generic market entry from select years between 1999 and 2019.

Let's pause for a moment and connect how these market trends relate to the valuation model. By now, you understand that LOE represents a critical inflection point in a drug's commercial lifecycle. It's the moment when your model needs to reflect a dramatic shift in two critical factors: how much a company can charge for each unit (pricing) and how many units they can sell (sales volume). Let's break down the typical patterns we from industry data and how you can incorporate them:

1. **Price:** When exclusivity ends and competitors enter the market, drug prices fall—often dramatically—with different drug types showing slightly different patterns of decline.

   ◆ **Small molecules** see prices fall by up to 50% within just one year after LOE (Conrad 2024).

   ◆ **Oral solid medications** follow a similar pattern, with average price declines of 40-50% in the first few years (Frank et al. 2021).

   ◆ **Biologics** experience a somewhat gentler decline—typically 25-55% within the first three years after LOE (Samsung Bioepis 2025, Association for Accessible Medicines 2024; Biosimilars Forum 2021)—likely because of slower biosimilar adoption, driven by incentives favoring brand-name products and greater barriers of market acceptance of "equivalent" manufacturing of these complex therapies

2. **Sales:** After LOE, brand-name drugs typically see a sharp decline in the number of units sold. Industry data reveals wide variances in these impacts across different drug types.

   ◆ **Small molecules** often lose up to 80% of their sales volume within the first year as patients switch to generics, as shown in Figure 24 (Grabowski et al. 2021).

   ◆ **Biologics** tend to hold their ground slightly better, losing around 20-70% of their sales volume within the first 3 years after first biosimilar entry (Samsung Bioepis 2025, Association for Accessible Medicines 2024; Goldman and Philipson 2021). This variability reflects the diverse competitive dynamics across biologic markets. Herceptin (trastuzumab) and Avastin (bevacizumab) experienced roughly 70% sales loss within three years post-LOE, whereas Humira (adalimumab) faced a projected 20-30% sales loss over the same period, illustrating markedly different biosimilar penetration patterns, as shown in Figure 25 (Samsung Bioepis 2025).

♦ Overall, biosimilar uptake remains slower than generic drug adoption, likely due to manufacturer incentive programs and market access strategies that help maintain brand sales. Samsung Bioepis (2025) provides valuable quarterly tracking of biosimilar launches and their impact on pricing and market dynamics.

BRAND MARKET SHARE RETENTION AFTER BIOSIMILAR ENTRY, 2015 TO 2025

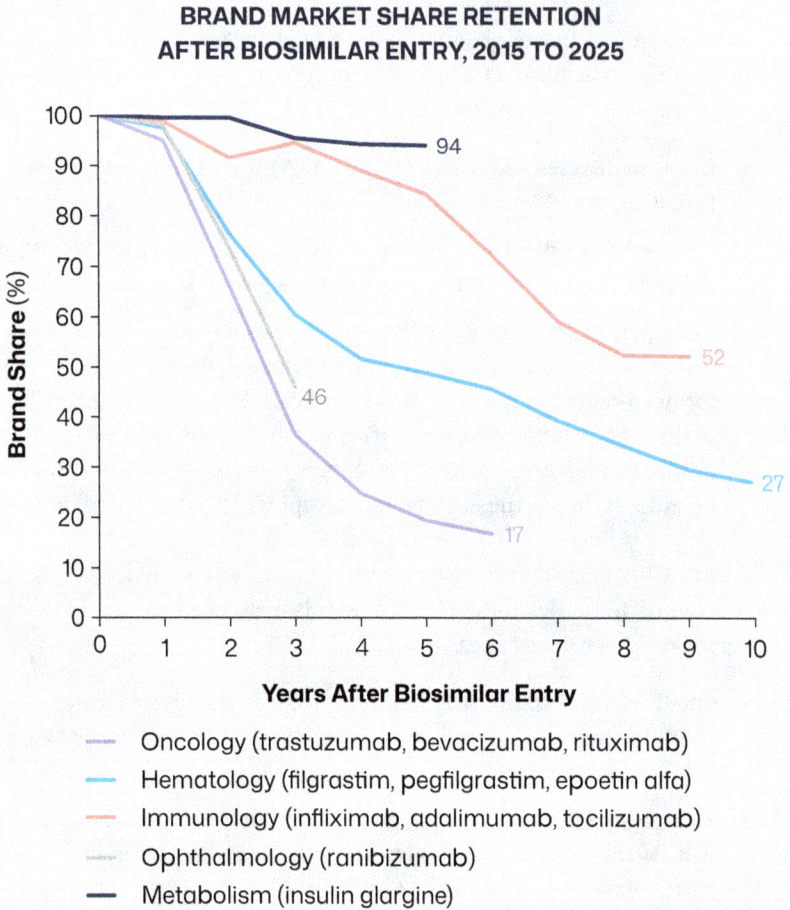

Oncology (trastuzumab, bevacizumab, rituximab)
Hematology (filgrastim, pegfilgrastim, epoetin alfa)
Immunology (infliximab, adalimumab, tocilizumab)
Ophthalmology (ranibizumab)
Metabolism (insulin glargine)

Note: Only one biological product is included in the therapeutic areas of ophthalmology and metabolism. Brand share is calculated as a product's sales volume divided by the total market sales volume, expressed as a percentage.

Analysis by AGMI Group based on data from Samsung Bioepis 2025

Figure 25. Yearly brand share performance for biologics by therapeutic area following the entry of biosimilars between 2015 and 2025.

Other factors to consider when modeling post-LOE revenue include the competitive landscape and brand strength. The number of generic competitors significantly impacts pricing erosion: a single competitor might only drive prices down by 30%, while five competitors could accelerate the decline to as much as 85% (U.S. Food and Drug Administration 2022). Brand loyalty can serve as a buffer, helping some drugs retain between 0–30% of their sales volume despite cheaper alternatives (PharmExec 2020). However, since brand loyalty is nearly impossible to predict for early-stage drug candidates still in development, it's generally best to take a conservative approach and not factor potential brand strength into your initial valuation models.

The key takeaway is that both prices and sales decline rapidly when market exclusivity ends and generic or biosimilars enter the market. Rather than calculating price and sales impact separately, we can use a simplified approach in our valuation model:

1. **Establish Peak Revenue Baseline:** Identify the maximum annual revenue achieved during the market exclusivity period (typically at the peak of the product uptake curve and last year of market exclusivity).

2. **Model Post-LOE Sales Decline:** Beginning in the year LOE occurs, apply a 50% year-over-year revenue decline.

3. **Reach the Market Share Floor:** Revenue stabilizes when the original product retains approximately 5–10% market share, representing loyal patients who remain on the branded product despite generic or biosimilar availability.

For example, if peak revenue is $800M (30% peak market share) and 5% is the market share floor, it would decrease to $400M (15% market share) in the first year post-LOE, $200M (7.5% market share) in the second year, and then stabilize at $100M (5% market share) in the following years. This approach effectively models price and sales erosion without unnecessary complexity. Since LOE is typically far off for most early-stage biotech products, this can serve as a reasonable approximation without introducing too much complexity.

> **Quick tip:** Remember that while these modeling guidelines provide a solid foundation, real-world LOE impacts vary considerably based on factors like therapeutic area, market structure, and competitive intensity. For most valuation work—especially for candidates approaching commercialization, the simplified peak revenue decline method we've outlined offers the right balance of practicality and accuracy. That said, certain situations may warrant deeper analysis: breakthrough therapies in novel areas, drugs with unique delivery mechanisms, or products facing an unusual competitive landscape may benefit from a more customized forecast that incorporates specific market research and competitive intelligence of price and sales to better predict post-LOE performance.

The Drug Commercialization Process

Commercialization is the process of bringing a new treatment from the lab to the hands of patients. Furthermore, it is also known as the market exclusivity period, which covers the years between launch year to the entry of generics or biosimilars. This stage involves setting up manufacturing pipelines, crafting marketing strategies, and building distribution networks to ensure the drug reaches healthcare providers and patients efficiently (Biotech Primer 2023; Goodarzi et al. 2017).

Manufacturers must obtain licenses, ramp up production, and launch campaigns with healthcare organizations, all while maintaining rigorous quality standards to guarantee safety and consistency. Even after a drug hits the market, ongoing monitoring ensures any potential safety concerns or side effects are quickly addressed. These processes of commercialization determine product success into the market and typically start once clinical trials are completed.

There are five steps to consider during the drug commercialization process as depicted in Figure 26 (Biotech Primer 2023):

01

Regulatory Approval

Seeking regulatory approval from the FDA involves submitting clinical trial data to prove the drug's safety and effectiveness. The agency will review the data and determine if the drug should be approved for the indicated use.

Manufacturing

02

After obtaining approval, the drug manufacturer must obtain a license to manufacture and distribute the drug. Contract Development and Manufacturing Organizations (CDMOs) come into play at this stage, assisting the pharmaceutical company in scaling the drug from lab to pilot and, if approved, to production scale. They help develop the active ingredient and formulation, and they also manufacture and package the drug for market sale. CDMOs play a vital partnership role, ensuring that the manufacturing process meets strict quality control standards, guaranteeing consistency and safety for every dose.

Marketing Strategy

03

The drug manufacturer will need a marketing strategy to promote the drug and make it available to healthcare providers and patients. This can involve advertising campaigns, sales representatives, and partnerships with healthcare organizations. The marketing strategy should also consider the target patient population, pricing, and distribution channels.

Launch

04

Once the manufacturing process is in place and the marketing strategy is developed, the drug can be launched and be made available through pharmacies, hospitals, and other healthcare providers. Furthermore, during the early stages of the drug launch, building stakeholder education and awareness, establishing patient support programs, and forming partnerships with key opinion leaders will be key to build momentum.

Post-Marketing Monitoring

05 Following the drug's launch, continuous monitoring and surveillance are crucial to spot any potential new safety issues or adverse effects. Drug producers must notify regulatory organizations of any adverse occurrences, and continuous monitoring can guarantee that patients receive safe and efficient therapies.

STEPS OF COMMERCIALIZATION

Manufacturing

Obtain a license to manufacture and distribute the product or drug.

Product Launch

Launch product once manufacturing and marketing strategy in place.

02

04

01

03

05

Regulatory Approval

Obtain regulatory approval from agencies such as the FDA.

Marketing Strategy

Develop a marketing strategy to promote the product and make it available to healthcare providers and patients.

Post-Marketing Monitoring

Monitor potential new safety issues or adverse effect drug post-launch with surveillance studies.

Adapted by AGMI Group from Biotech Primer 2023

Figure 26. Overview of key steps in the drug commercialization process.

However, several critical areas can hinder successful drug commercialization. Here are three key challenges to consider during the process:

1. **Poor Communication and Organizational Silos:** Effective communication is critical between different organization structures and partnerships. Many initiatives and drug launches come up short because they lack the communication skills for clear, consistent messaging necessary to motivate, educate, and foster product commitment.

    a. A recent poll revealed that only 11% of respondents felt that insights meaningfully influence strategy within their organizations (Within3 2024). To accelerate drug commercialization, it is vital to surface insights more quickly and frequently, particularly by breaking organizational and data silos, identified by 39% of respondents as the primary bottleneck to insight generation. Effectively communicating the value of these insights to senior leadership is crucial for driving strategic impact.

2. **Lack of Clear Unique Selling and Value Propositions:** Many drug launches fail because they don't effectively communicate their core value. This often stems from limited engagement with patients, physicians, and other key stakeholders. A well-defined patient journey, highlighting unmet needs, treatment challenges, and points of friction in care, can shape a compelling narrative that resonates with the market. By addressing patient needs, treatment challenges, and stakeholder concerns, companies can enhance market perception and drive adoption.

3. **Poor Competitor Understanding:** Market and competitor research are essential for shaping an effective commercialization strategy. Learning from competitors' successes and failures can help refine your approach. However, many pharmaceutical companies fail to implement the necessary work or technologies to accurately track and analyze competitor behavior and activities, resulting in missed opportunities to improve their own processes.

4. **Inefficient Supply Chain Management:** Bringing a pharmaceutical product to market requires a careful balance between efficiency and quality. Effective supply chain management encompasses four major areas: packaging and labeling, inventory management and transportation, product distribution, and tracking and tracing. Any misstep in these areas can lead to delays, increased costs, or regulatory setbacks.

    a. To mitigate these risks, companies must invest in demand management strategies and leverage available technologies to optimize logistics, improve forecasting accuracy, and ensure compliance with evolving regulations.

If you're interested in exploring the drug commercialization process in more detail, several helpful guides and drug launch checklists are available from sources like Censora (2025), Within3 (2024), and BioSpace (2021).

Cost of Goods Sold (COGS) Fundamentals

The COGS encompasses the direct costs of producing the goods sold by a company or product. This amount includes the cost of the materials, labor, and manufacturing overhead that is directly used to create the goods. It excludes indirect expenses, such as administrative, marketing and sales, and R&D costs. Generally, factors that contribute to COGS include:

- **Manufacturing Costs:** Costs incurred during the production of the drug, including labor, materials, and overhead.
- **Packaging and Distribution:** Costs related to packaging the drug and distributing it to wholesalers, pharmacies, and healthcare providers.
- **Quality Control:** Expenses associated with ensuring the drug meets regulatory and quality standards.
- **Scalability:** As production volumes increase, economies of scale can reduce COGS, improving margins.

Many biotech and pharma companies often lack a comprehensive understanding of what drives COGS, which limits their ability to manage these expenses effectively. As inflation, regulatory changes, and the Inflation Reduction Act of 2022 take effect, COGS is anticipated to increase over the next 3 to 5 years—making it increasingly important to understand the underlying processes that contribute to these costs (Celen et al. 2023).

COGS is usually reported as a percentage of gross revenue and varies widely depending on production costs and pricing. To clarify, whenever we refer to COGS, we will refer to it as a percentage of revenue and not the direct costs. The weighted average COGS for the pharma industry globally in 2021 was approximately 26% of revenue (Pierre 2022).

Our analysis at AGMI Group examined 170 public brand-name companies with market capitalizations below \$2B using EvaluatePharma data. The study revealed average COGS of 16% for small molecules, 10% for biologics, and 30–35% for cell and gene therapies.

These variations in COGS stem from two primary factors:

1. **Production Costs:** Average production varies by complexity depending on the biotech product being manufactured and distributed. A 2017 analysis conducted by Boston Consulting Group showed that small molecules have a production cost of \$5 per pack compared to biologics which have a production cost \$60 per pack on average (Gooch et al. 2017). As a result, biotech companies with a larger share of biologics in their portfolio tend to have higher average costs per pack, which can significantly influence the overall cost performance of the company.

2. **Product Pricing:** Because it is reported as a percentage of gross revenue, COGS will depend on the price being charged. It is well known that biotech companies have a lower COGS on average than the S&P 500 companies because of the higher price being charged for the biotech products (Ledley et al. 2020).

    ♦ For example, COGS can differ greatly between brand-name/innovative and generic biotech companies as seen in Figure 27 (Positano et al. 2019). The median COGS for generic business units ranged from 25–42%, significantly higher than the 19% observed for high-end companies. Additionally, when further categorized, generic companies can be divided into biosimilars and small molecules. This analysis reveals that biosimilars have a lower median COGS of 25%, compared to 42% for generics. These numbers are similar to previous cost analysis by company size and type in 2008 (Ledley et al. 2020).

    ♦ The higher COGS for generics stems from their significantly lower retail prices compared to brand-name small molecules, biologics, and even biosimilars. On average, biologics prices can be approximately 22 times more than small-molecule drugs, while biosimilar entry can reduce the price of biologics by up to 56% (Goldman and Philipson 2021). Because generics are sold at lower prices, their revenue is reduced, making production costs a larger percentage of the revenue. This highlights the profound impact of pricing on COGS relative to production costs.

COGS AS PERCENTAGE OF REVENUE BY COMPANY TYPE AND GENERIC BUSINESS UNITS, 2019

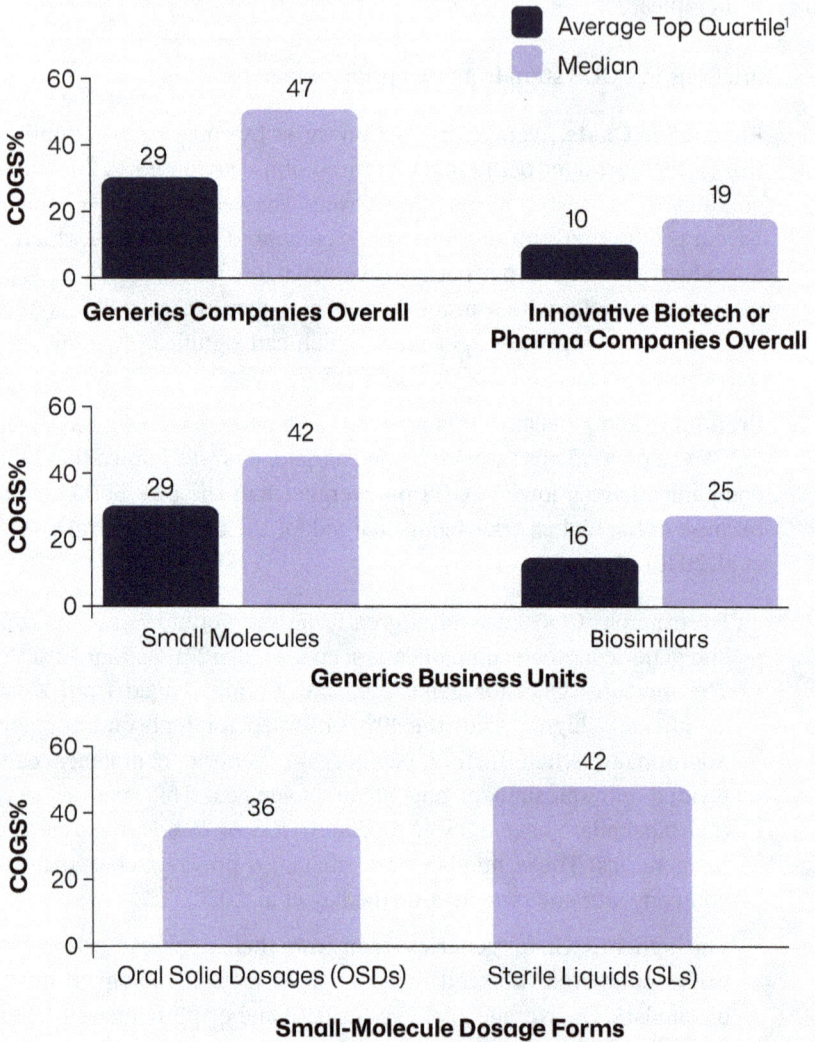

Figure 27. Comparison of COGS as a percentage of revenue across different pharmaceutical company types and generic business units for 2019.

For our valuation of early-stage biotech companies, we recommend estimating based on pricing as it is a clear and large driver of COGS. In general, for highly-priced products (e.g., biologics), we typically estimate COGS to be 10–20% of revenue while lower-priced products (e.g., small-molecules in primary care indications) can have COGS up to 25–35% of revenue (Positano et al. 2019; Gooch et al. 2017; Jiang et al. 2022; Ledley et al. 2020). Based on our internal analysis, we recommend applying higher COGS estimates of 30-35% of revenue (up to 50%) for cell and gene therapies, reflecting the increased manufacturing complexity and production costs relative to traditional biologics (Callens et al. 2016; Macdonald 2023).

For clarity, this applies only during the sales or revenue generating period. We typically assume a constant COGS value throughout the commercialization process. While COGS may fluctuate from year to year, it is reasonable to use a constant value for early-stage products. Additionally, any initial manufacturing development activities during the pre-launch phase are generally categorized as R&D expenses rather than COGS.

Effectively managing these costs is essential for maintaining a company's competitive edge. Success depends on factors such as portfolio management, operational efficiencies, and strategic manufacturing networks or partnerships. However, as previously mentioned, manufacturing costs are rising rapidly due to three key factors: (1) the emergence of novel drug modalities and technologies targeting smaller addressable patient populations, (2) the shift toward more personalized supply chains, and (3) the growing emphasis on supply chain resilience in response to global disruptions, particularly inflation-driven increases in raw material and distribution costs in the post-COVID-19 era.

Figure 28 illustrates the inverse relationship between the median addressable U.S. patient population for the top ten biotech products by sales and the percentage of drug sales targeting fewer than 200,000 patients relative to global prescription spending (Celen et al. 2023).

ADDRESSABLE POPULATION OF TOP 10 DRUGS AND SALES OF DRUGS TARGETING <200K PATIENTS, 2000 TO 2025

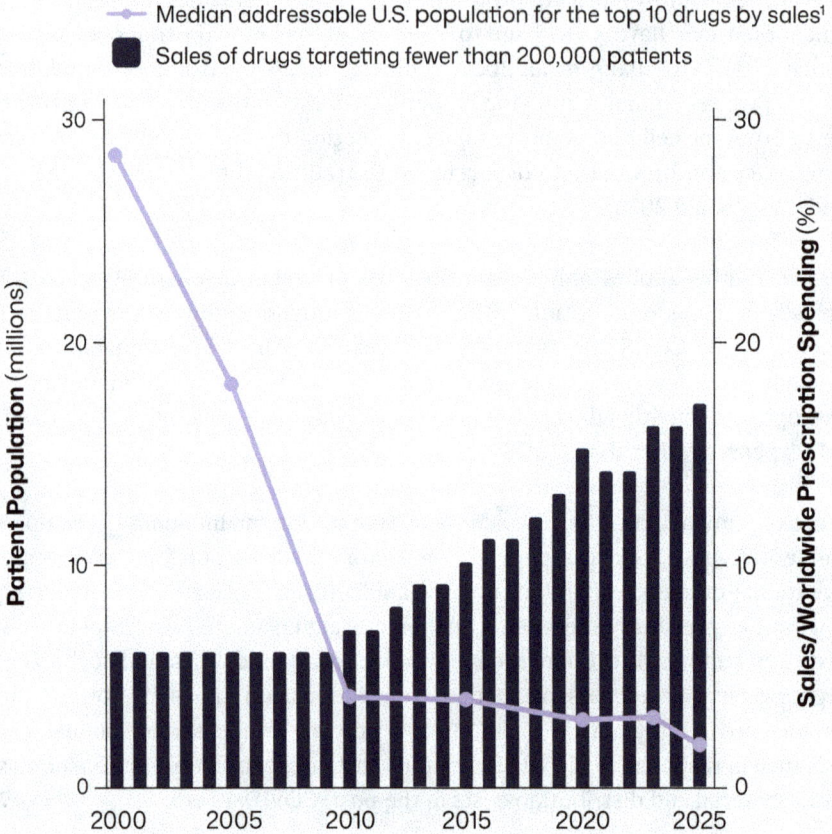

Figure 28. Addressable patient population of the top 10 drugs and drugs sales targeting fewer than 200k patients from 2000 to 2025.

[1] Addressable population was calculated using prevalence of first or major indication that was marketed. If prevalence data was not available, incidence rates were used instead.

Note: Sales data is from 2000 to 2022. Sales if Pfizer's pneumococcal vaccine Prevnar 13 and COVID-19 vaccines + treatments were excluded from the analysis. Years 2023 to 2025 are forecasted sales.

Source: Celen et al. 2023

In light of increasing costs, there are several strategies to manage COGS effectively. Companies can approach this in a few different ways, ranging from small changes in specific areas to more comprehensive plans to cut costs. Here are four example strategies to help reduce these costs:

1. **Improve Internal and External Manufacturing Networks:** Companies need to ensure that their internal and external manufacturing networks are optimized to support the demand and capacity profile of each modality over the long term, reevaluating their manufacturing network strategy every 5 years to ensure that it is still optimized.

2. **Increase Flexibility in Manufacturing:** By building modular manufacturing plants, companies can scale up production only for products that have received regulatory approval, instead of building capacity before it's necessary. Companies that invest in manufacturing flexibility can stagger their investment decisions and make site and equipment purchases as much as 2–4 years later compared to traditional manufacturing setups. For example, Sanofi completed building its $554M pandemic-ready modular plants that are designed to produce multiple vaccines and biologics platforms (Dunleavy 2024).

3. **Focus on Operational Efficiency:** Companies can adopt methods that help them work more effectively, reducing waste and improving productivity across the manufacturing network, and address bottlenecks. Investing in data analytics across the value chain to promote smarter decisions can potentially increase productivity by 6–7% each year (Celen et al. 2023).

4. **Streamline Decision-Making:** Organizations should simplify their structures to make decision-making quicker and more effective. This includes cutting down on unnecessary management layers and overlapping roles, which helps eliminate tasks that don't add much value.

To sustain long-term competitiveness, biotech companies must effectively manage COGS across both their legacy portfolios and innovative pipelines, which involves striking a balance between product innovation, supply chain resilience, and production costs.

What to Keep in Mind

Think of launch curves, exclusivity periods, and commercialization strategies as your treasure map to a drug's financial success. You've seen how these elements work together to tell a story about value—from the moment doctors first prescribe a new therapy to the years it enjoys patent protection. Remember, even the most promising molecule isn't worth much without thoughtful planning around adoption, exclusivity timing, and manufacturing efficiency. These aren't just theoretical concepts—they're the real-world drivers that generate those all-important revenue streams to recoup your development investments.

Now that you've got these market dynamics in your toolkit, we're ready to step back in time to where it all begins: the R&D journey. In Chapter 4, we'll walk through the winding path of drug development—exploring how long each stage typically takes, what it costs, and your chances of success along the way. By understanding both where you're going (commercialization) and how you'll get there (R&D), you'll have everything you need to build valuation models that actually make sense in the complex world of biotech.

BEFORE YOU WRAP UP

Take a moment to reflect on what you've learned about product dynamics. These key takeaways and questions are designed to help you connect these concepts to your biotech valuation model and identify areas you'd like to explore further.

Key Takeaways

- A drug's commercial success is modeled through launch curves, with two key variables: time to peak (averaging 8 years) that varies with therapeutic area and the S-shaped curve that can be shifted left or right depending on market uptake factors.

- Loss of exclusivity (LOE) triggers dramatic revenue declines—small molecules typically lose 80% of sales volume within a year, while biologics experience a somewhat less severe but still significant drop (22–50%).

- Product differentiation, unmet clinical need, and pricing strategy directly influence both the shape of the launch curve and time to peak sales, making these factors critical in accurate revenue projections.

- Cost of Goods Sold (COGS) varies significantly by product type—biologics typically maintain better profit margins (10–20% of revenue) compared to small molecules (25–35% of revenue) due to higher pricing despite higher manufacturing costs.

- Successful commercialization involves much more than regulatory approval, requiring manufacturing scale-up, marketing strategy, distribution networks, and post-marketing monitoring to translate clinical success into market performance.

BEFORE YOU WRAP UP *(Continued)*

Reflection Questions

1. How might the launch curve for your product differ from the standard 8-year time to peak based on your therapeutic area, level of innovation, and competitive landscape?

2. Which factors might accelerate or delay market uptake for your specific product? Consider pricing strategy, unmet needs, therapeutic benefit, and competitive differentiation.

3. How would your valuation change if your product reached peak sales in 6 years versus 10 years?

4. What post-LOE strategy might help maintain some market share against generic or biosimilar competition? Consider product differentiation, patient support programs, or line extensions.

5. How might your COGS projections differ based on your product type, pricing strategy, and manufacturing approach?

6. Which commercialization challenges do you anticipate being most significant for your specific product, and what strategies might help overcome them?

04

Research and Development

I magine spending billions of dollars and over a decade of work, only to have your new biotech product fail just before it reaches the market. This is the harsh reality of drug development in the biotech industry. The biggest costs in creating a new drug are tied to the time, resources, and high risks involved. At the heart of this process is research and development (R&D), which transforms early-stage discoveries into clinical products ready for market. However, this journey demands substantial upfront investment, often with no guarantee of success.

As you continue shaping your biotech forecast, this chapter will help you discover the intricate path of R&D from discovery to approval, exploring the time, costs, and critical decision points that make (or break) a drug's chance of success. Whether you're a founder evaluating your startup's runway, an investor calculating risk-adjusted returns, or a biotech professional planning development timelines, understanding these R&D fundamentals will transform how you approach valuation. You'll learn to identify which therapeutic areas and modalities offer higher success rates, how to estimate realistic development costs and time for your specific indication, and most importantly, how to incorporate probability of success into your financial projections. By mastering these concepts, you'll be equipped to make more informed investment decisions, create compelling business cases, and build valuation models that accurately reflect the high-risk, high-reward nature of bringing new therapeutics to market.

What Predicts Drug Development Success

With the immense cost and time required for R&D, understanding what drives a drug from Phase 1 trials to approval is critical. QLS Advisors took on this challenge, conducting a deep analysis of the drug development landscape from 2011 to 2020, as seen in Figure 29 (QLS Advisors 2021). They identified "features" such as the characteristics of the drug, its indication, sponsor/company, and clinical trial design to identify what signals contribute to drug development success. Using a random forest classifier, they pinpointed the most influential factors that predict success, offering a fascinating glimpse into the patterns that can make or break a drug's journey to market.

FEATURE IMPORTANCE BY DEVELOPMENT PHASE TO APPROVAL

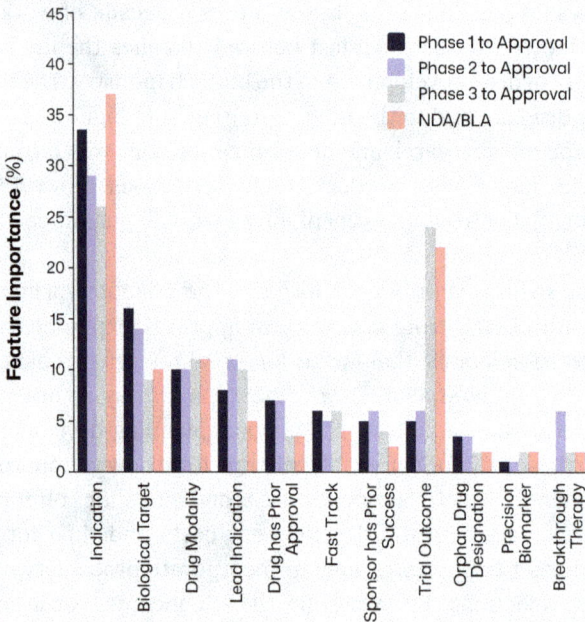

Note: Random forest classifier was used across all predictions. Individual predictions may have unique drivers specific to a given therapeutic area, trial design, or drug profile that are not listed here.

Source: QLS Advisors 2021

Figure 29. Most important features by clinical trial phases to approval using a random forest classifier. Feature importance is normalized to sum to 100%.

In their analysis, therapeutic indication, a concept we previously discussed in Chapter 1, was found to be the consistently ranked top feature across all clinical development phases. The report notes that there is significant variability across indications even within a therapeutic area. For example, in oncology, the Phase 1 to approval rate ranged from 1% for pancreatic cancer to 15% for gastrointestinal cancers. We will discuss more about success rates by therapeutic area in the next section.

Other interesting features to consider include:

1. **Lead Indication:** A drug might have better overall success if the sponsor initiates multiple clinical trials for different indications that the drug wasn't originally designed to treat. In other words, by testing the drug in a wide range of potential uses, the sponsor increases the chances that at least one trial will lead to a successful outcome, even if many others fail.

2. **Trial Outcome:** A drug-indication pair whose trial failed to meet its endpoints has a low probability of success in advancing to approval.

3. **Prior Approval of a Drug for Another Indication:** Developing an already approved drug for a new indication has a higher chance of success, particularly when the manufacturing process is already well established and the drug has been proven safe for patient use.

4. **Use of Pre-Selection Biomarkers:** Use of targeted agents and, therefore, patient pre-selection biomarkers in clinical trials has become more common. For example, between 2016 and 2020, 85% of oncology approvals were for targeted agents that utilized pre-selection against a driver mutation or targeted a tumor-specific antigen (QLS Advisors 2021). Another study examining clinical data supports using biomarkers in clinical trial designs to increase the probability of success (Wong et al. 2019).

5. **Drug Modality Type**

    a. Monoclonal antibodies have demonstrated higher success rates from Phase 1 to approval compared to other treatment modalities. In oncology, immunotherapies targeting PD-1 or PD-L1 have been particularly successful, with 24% advancing from Phase 1 (QLS Advisors 2021). In contrast, other immunotherapies have been much more likely to result in failure from Phase 1 (2%).

6. **Breakthrough Therapy**

    a. A drug may offer significant improvement in treatment outcomes compared to existing therapies and would be given the Breakthrough Therapy Designation by the FDA for faster approval.

7. **Sponsor Track Record**

    a. Successful phase transitions are likely associated with operational experience of the sponsor or company given their experience conducting clinical trials and navigating the regulatory review process.

Success Rates by Therapeutic Area

Therapeutic indication remains the most critical factor influencing the successful progression from clinical trials to approval. However, even within the same therapeutic area, the probability of success from Phase 1 to approval varies significantly depending on the specific indication. To provide a broader perspective, QLS Advisors reported probability of success from Phase 1 to approval by therapeutic area, highlighting which areas tend to have more successful indications on average, as shown in Table 6 (QLS Advisors 2021).

PHASE 1 TO APPROVAL SUCCESS RATES BY
THERAPEUTIC AREA, 2011 TO 2020

Therapeutic Area	Success Rates Phase 1 to Approval	N
Hematology	23.9%	352
Metabolic	15.5%	399
Infectious disease	13.2%	1170
Others	13.0%	541
Ophthalmology	11.9%	415
Autoimmune	10.7%	1305
Allergy	10.3%	201
Gastroenterology	8.3%	186
All indications	**7.9%**	**12728**
Respiratory	7.5%	501
Psychiatry	7.3%	442
Endocrine	6.6%	887
Neurology	5.9%	1411
Oncology	5.3%	4179
Cardiovascular	4.8%	651
Urology	3.6%	88

Source: QLS Advisors 2021

Table 6.   Analysis of drug development success rates from Phase 1 to approval, organized by therapeutic area from 2011 to 2020.

Success rates from Phase 1 to approval vary significantly across different therapeutic areas. For example, hematology leads with the highest success rate at 23.9%, followed by metabolic diseases at 15.5% and infectious diseases at 13.2%. Four other disease areas—ophthalmology, autoimmune, allergy, and gastroenterology—exceed the overall average success rate of 7.9%, with rates ranging from 8.3%–11.9%. Just below the average are respiratory diseases at 7.5%. Notably, five disease categories fall well below the average: endocrine, neurology, oncology, cardiovascular, and urology. Note that both oncology and neurology, with their large N values, likely driven from clinical need and market demand, significantly contribute to the overall industry's lower approval rate given their low success rates. Rare disease therapies were notably successful with an overall LOA of 17.0% versus Chronic, high prevalence disease therapies were less successful with an overall LOA of 5.9%.

> **Quick tip:** Note that success probabilities fluctuate annually, so it's essential to access the most current published analysis.

Success Rates by Therapeutic Modality

The therapeutic modality a drug is built upon plays a significant role in determining its probability of success. As shown in Table 7, data from QLS Advisors highlights the variability in success rates from Phase 1 to approval across different modalities (QLS Advisors 2021).

PHASE 1 TO APPROVAL SUCCESS RATES BY THERAPEUTIC MODALITY, 2011 TO 2020

	Success Rates	
Therapeutic Modality	**Phase 1 to Approval**	**N**
CAR-T	17.3%	67
siRNA/RNAi	13.5%	87
Monoclonal antibody	12.1%	2136

ADCs	10.8%	184
Gene therapy	10.0%	96
Vaccine	9.7%	316
Protein	9.4%	800
Peptide	8.0%	619
Small molecule	7.5%	7171
Antisense	5.2%	162

Source: QLS Advisors 2021

Table 7. *Analysis of drug development success rates from Phase 1 to approval, organized by therapeutic modality from 2011 to 2020.*

Chimeric antigen receptor T-cell (CAR-T) therapies achieved the highest success rate, at 17.3%, followed by RNA interference (RNAi) therapies at 13.5% and monoclonal antibodies at 12.1%. Notably, small molecules showed a lower overall success rate of 7.5%. One way to interpret this is that more biologically complex modalities may be associated with higher clinical success, though this relationship is not always linear.

It's important to note that this dataset predates the widespread adoption of RNA-based therapeutics following the COVID-19 pandemic, which may have shifted some success trajectories in recent years.

Other highlights from the QLS Advisors report include:

1. Phase 2 remains the largest hurdle in drug development with just 28.9% of candidates reaching this phase transition.

2. Immuno-oncology therapies show promise, with a success rate of 12.4%, compared to just 5.3% for other oncology approaches.

3. Rare disease therapies had higher success rates, at 17.0%, compared to treatments for common, chronic conditions, which had a success rate of 5.9%.

4. Use of preselection biomarkers during the drug development process doubled the probability of success.

Overview of the Drug Development Stages

Drug discovery and development is a long, expensive, and high-risk process—typically taking 10–15 years and costing $1–2B per approved drug (Hinkson et al. 2020). One of the first major challenges is optimizing drug candidates at the Preclinical Stage for Phase 1 clinical trials. However, the path ahead is filled with uncertainty: around 90% of drugs fail during the clinical trial phases or in the approval process (Dowden and Munro 2019; Takebe et al. 2018). When preclinical candidates are included, the overall failure rate climbs even higher.

For you as a founder, investor, or biotech operator, understanding where and why drug candidates fail is essential at this stage of your journey. You might be wondering, "How do I accurately value a pre-clinical asset?" or "What are realistic odds my Phase 2 candidate reaches the market?" This knowledge empowers you to ask better questions in due diligence, make informed go/no-go decisions at critical development milestones, and build more realistic valuation models that won't collapse when faced with setbacks. Rather than being discouraged by these failure rates, you'll learn to use them as powerful calibration tools—transforming optimistic projections into probability-weighted forecasts that experienced investors and partners take seriously.

As we explore the specific reasons for failure at each development stage, you'll gain insights that directly translate to smarter capital allocation and more accurate financial modeling—key skills we'll build on in subsequent chapters when constructing complete valuation frameworks.

Why do drugs fail during clinical trials?

Drug development follows a vigorous process of genetic and genomic target validation, screening of drug candidates, optimization of drug activity and properties, preclinical and toxicity testing, biomarker selection, and clinical trial designs. The 90% failure rate to get through clinical trials have been attributed to (Sun et al. 2022):

1. **Lack of Clinical Efficacy (40–50%):** Biological discrepancy among in vitro, in vivo animal models and human pathophysiology, resulting in inconsistent outcomes.

2. **Unmanageable Toxicity (30%):** Off-target or on-target effects; toxic drug accumulation in vital organs or blood cells.

3. **Poor Drug-Like Properties (10–15%):** Formulation, protein binding, metabolic stability, or in vivo pharmacokinetics.

4. **Lack of Commercial Needs and Poor Strategic Planning (10%):** Change in therapeutic focus, market changes, company mergers, or poor conduction of clinical studies.

Furthermore, each phase in drug development can be vastly different from each other. Figure 30 provides an overview of different phases, from Discovery and Preclinical Stage (target validation, drug candidate screening, lead optimization, and preclinical testing) to clinical trials in humans (Phase 1 through 3), followed by approval. Each phase varies in duration, cost, and probability of success (Sertkaya et al. 2024; QLS Advisors 2021; Schlander et al. 2021; Paul et al. 2010).

DRUG DEVELOPMENT TIMELINE OVERVIEW

Candidates	Stage	Time Duration	Probability of Success	% of Total Cost*	Description
	Target Validation	~1 year	80%	1%	Disease models, Target identification, Target validation.
>10,000	Compound Screening	~1.5 years	75%	2%	Visual screening, HTS (High-Throughput Screening).
~250	Lead Optimization	~2 years	85%	8%	SAR (Structure-Activity Relationship), Drug-like properties, Solubility, Permeability,
10-20	Preclinical Test	~1 year	69%	4%	ADME (Absorption, Distribution, Metabolism, Excretion), Plasma PK, Efficacy, Toxicity.
~6	Phase 1	~2.3 years	60%	5%	PK (Pharmacokinetics), Dose escalation, Toxicity.
~4	Phase 2	~3.6 years	36%	15%	Dose optimization, Efficacy proof-of-concept.
~2	Phase 3	~3.3 years	66%	63%	Large-scale efficacy, Safety confirmation.
1	Approval for Launch	~1.3 years	88%	2%	Regulatory review, Market approval process.

Total cost percentage assumes multiple clinical trials for Phase 1, 2, and 3

Source: Sertkaya et al. 2024, QLS Advisors 2021, Schlander et al. 2021, and Paul et al. 2010

Figure 30. *Drug development pipeline illustrating the length of time, success rates, and cost for each phase from discovery to approval.*

Now, let's go over each of the phases in more detail.

Discovery and Preclinical Stage

The Discovery and Preclinical Stage is the critical first step in drug development, where researchers identify promising new treatments and ensure their safety before formal testing in humans. This phase consists of two main parts: the Discovery Phase and the Preclinical Stage, with estimated costs ranging from $15–$100M (Roden 2023).

A systematic review of research published up to 2020 found that these two phases typically span 4–7 years (BioStock 2023). Some studies suggest that the cost can be up to one-third of total drug development costs (Schlander et al. 2021).

While data on success rates in the Discovery Phase is limited, the probability of a candidate advancing from discovery to preclinical testing to be around 51% (Paul et al. 2010). In the Preclinical Stage, there is much more data available and the average probability of success among studies is estimated to be around 69% (Schlander et al. 2021; Paul et al. 2010). Overall, the combined success rate for progressing from discovery through preclinical testing is approximately 35%.

Keep reading for a more detailed breakdown of the costs, duration, and probabilities of success to help you obtain a clearer picture of their components of each phase.

Discovery Phase

The Discovery Phase involves several key steps aimed at finding and optimizing potential drug candidates:

1. **Target Validation:** This initial step involves identifying and validating a biological target using a large library of chemical compounds or biological molecules to find whether the target has the desired effect. This stage typically takes one year, costs around $1.5M (2025 USD), and has around an 80% probability of success.

2. **Compound Screening**: Once promising biological targets and compounds are identified, the drug candidates undergo further screening and refinement to find "lead" compounds. This stage takes about 1.5 years, costs around $3.7M (2025 USD), and has a 75% probability of success.

3. **Lead Optimization**: In this step, the lead compounds undergo chemical modifications to enhance their properties. Researchers focus on improving the compound's potency, reducing potential side effects, and increasing its stability. This process typically takes 1.5 years, costs $14.9M (2025 USD), and has an 85% probability of success.

Preclinical Stage

The Preclinical Stage is designed to assess the safety and biological activity of the optimized lead compounds before they can be tested in humans, usually in animal models. This stage takes about one year, costs approximately $7.5M (2025 USD), and has a 69% probability of success. This phase includes several critical components:

1. **Toxicology Studies**: These studies evaluate the potential toxicity of the drug candidate. Researchers conduct tests on cell cultures and animal models to determine the compound's safety profile, including any adverse effects on organs and systems. The goal is to identify a safe starting dose for human trials.

2. **Pharmacokinetic and Pharmacodynamic Studies**: Pharmacokinetic studies investigate how the drug is absorbed, distributed, metabolized, and excreted in the body. Pharmacodynamic studies assess the drug's biological effects and its mechanism of action. Together, these studies provide crucial information about the drug's behavior in the body and its potential therapeutic effects.

3. **Formulation Development**: Researchers work on developing a suitable formulation for the drug, determining the best method of delivery (e.g., oral, intravenous) and ensuring that the drug remains stable and effective throughout its shelf life.

4. **Investigational New Drug (IND) Application**: Before clinical trials can begin, researchers must submit an IND application to the FDA. This application includes data from preclinical studies and outlines the proposed plan for human trials. Approval of the IND allows the transition from preclinical testing to clinical development.

Table 8 provides a summary of the duration, cost, and probability of success in a condensed chart for your reference (Schlander et al. 2021; Paul et al. 2010).

**DISCOVERY AND PRECLINICAL STAGE:
TIME DURATION, COST, & PROBABILITY OF SUCCESS**

		Duration	Cost (2025 $M)	Probability of Success
Discovery	Target Validation	1 year	$1.5	80%
	Compound Screening	1.5 years	$3.7	75%
	Lead Optimization	2 years	$14.9	85%
Preclinical		1 year	$7.5	69%
	Total	**5.5 years** (range 4-7 years)	**$28** (range $15-$100)	**35%**

Source: Schlander et al. 2021 and Paul et al. 2010

Table 8.   Duration, cost, and probability of success metrics for the Discovery and Preclinical Stage.

Clinical Trials

The clinical trial phase marks a crucial step in drug development, where experimental therapies are tested on human participants to evaluate their safety and efficacy. Each phase of a clinical trial builds on the last and serves a distinct purpose:

01   **Phase 1** involves initial testing on a small group to evaluate the treatment's safety and identify potential side effects.

02   **Phase 2** expands testing to a larger group, measuring the treatment's efficacy while gaining a more detailed understanding of its safety.

03   **Phase 3** then compares the experimental therapy with current standard therapies to confirm its efficacy and safety, ensuring it meets regulatory standards for widespread patient use.

As discussed earlier, time and cost are among the biggest challenges in R&D, as they are deeply interconnected. Our biotech valuation model incorporates key factors—trial duration, costs, and probability of success—to more accurately simulate the clinical development process. By applying risk adjustments, we ensure that our projections align with real-world outcomes.

In the following sections, we will dive deeper into trial duration, costs, and probability of success, exploring how each factor influences drug development and valuation.

Duration of Drug Development

Time directly impacts cost: the longer the development process, the higher the expenses. According to a study by QLS Advisors, the average time from Phase 1 trials to approval was 10.5 years between 2011 and 2020—excluding non-clinical factors such as internal decision-making and strategic execution (Paul et al. 2010). Figure 31 showcases a breakdown of the timeline by therapeutic area. Notice there are great variations in clinical phase durations (QLS Advisors 2021). For example, oncology has the longest Phase 1 transition at 2.7 years, while its Phase 2 and 3 durations are closer to the overall average.

PHASE TRANSITION DURATIONS BY THERAPEUTIC AREA

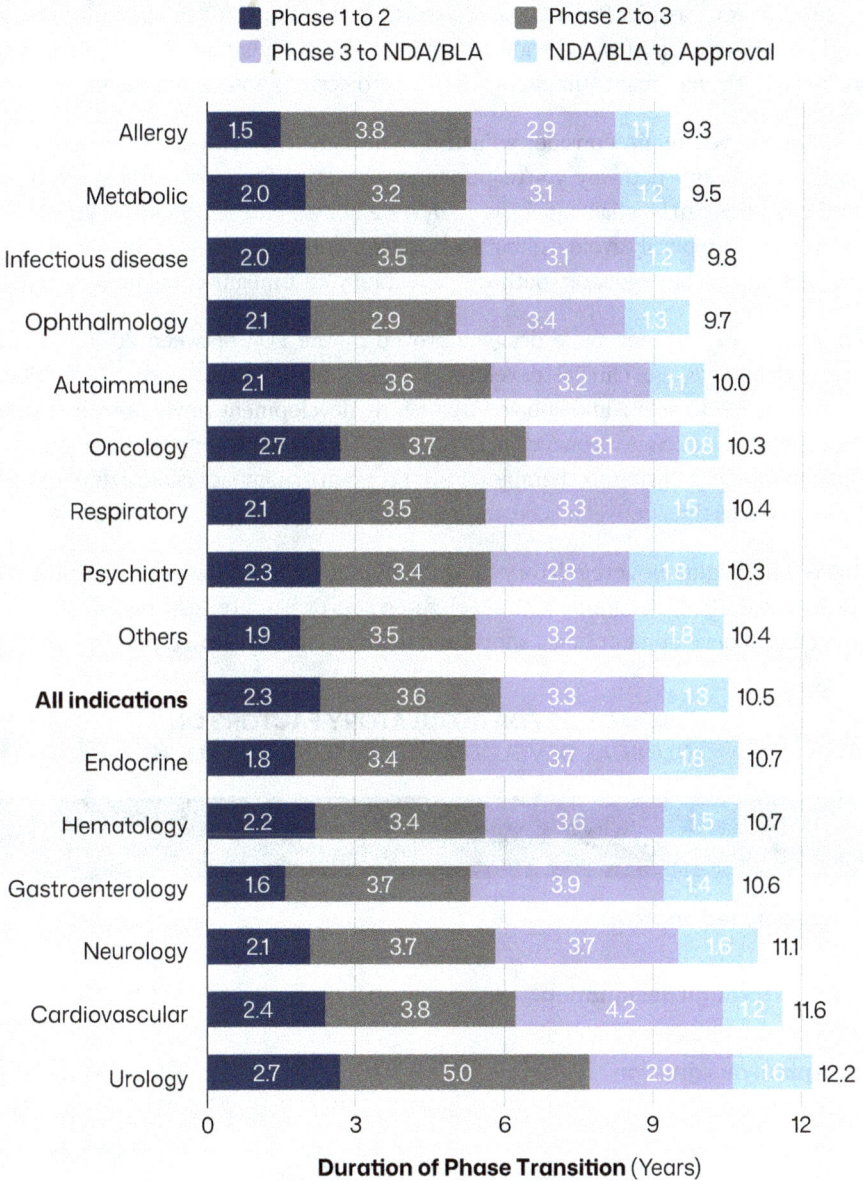

Figure 31. *Average duration for phase transitions during clinical development to approval.*

Source: QLS Advisors 2021

For Phase 1 trials, we use the average duration without adjusting for the therapeutic area, as these trials are relatively short and have a narrower duration range (Wong et al. 2019). In contrast, for Phase 2 and Phase 3 trials, we adjust for therapeutic area due to their longer durations and wider variability. These adjustments are more crucial, as trial length significantly impacts financial projections and biotech valuation.

Additionally, you might consider adjusting your clinical development timeline based on the regulatory pathway you're pursuing. However, for most early-stage drugs, predicting this can be challenging, as many have not yet entered clinical development. As a result, we typically avoid making such adjustments at this stage. That said, if you're fairly certain of the regulatory pathway, adjusting your timeline could be beneficial.

An analysis of 440 innovative drugs approved by the FDA between 2010 and 2020 revealed an average clinical development time of 9.1 years, with a confidence interval of 8.2–10 years, and showed that yearly development times remained stable throughout the decade (Brown et al. 2022). The FDA has many programs that facilitate the development of various therapies. Each program can impact clinical development time as well as time between licensing application submission and approval.

Table 9 highlights how regulatory factors influence clinical development timelines (Brown et al. 2022). Programs with accelerated approval or breakthrough designation saw clinical development times shortened by three or 1.3 years, respectively.

IMPACT OF FDA REGULATORY FACTORS ON CLINICAL DEVELOPMENT TIMELINE DURATION

Regulatory Factors	Effect (Years)	95% CI
Accelerated approval	–3.0	(–4.5, –1.5)
Breakthrough designation	–1.3	(–2.6, 0.0)
Orphan designation	1.5	(+0.4, +2.6)
> 1 Review cycle	1.8	(+0.4, +3.2)

Source: Brown et al. 2022

Table 9. *Analysis showing the impact of various FDA regulatory factors on clinical development timeline duration.*

In contrast, orphan designation was associated with an increase in development time by 1.5 years, despite smaller trial sizes. This delay may stem from challenges such as patient recruitment, limited understanding of the disease's natural history, and the need for novel clinical endpoints. However, orphan products benefit from an additional 2 years of marketing exclusivity in the U.S., which partially offsets these extended clinical development timelines and costs.

As expected, products requiring more than one review cycle (e.g., failures to win approval within the first review cycle) experienced longer development times by 1.8 years. Priority review status was not significantly correlated with overall clinical development time, likely due to limited statistical power in the analysis. Fast-track designation was not correlated with development times.

Based on the previous studies, we use these base assumptions when estimating the time duration of each clinical trial phase:

- **Phase 1:** 2.3 years
- **Phase 2:** 3.6 years (adjust by therapeutic area or specific program)
- **Phase 3:** 3.3 years (adjust by therapeutic area or specific program)

Cost of Drug Development

Estimating the cost of drug development can be difficult as different sources report their analysis based on pivotal trials, approved drugs on the market, aggregate per-phase data that do not report the number of trials executed, and exclusion of costs such as infrastructure cost, legal fees, etc.

For forecasting in biotech, the most practical approach is to apply a per-patient cost estimate to the projected study size. It's also important to consider that many biotech companies conduct multiple trials simultaneously.

A 2024 study analyzing drug R&D from 2000 to 2018—using data sources from ClinicalTrials.gov and custom tabulations from the FDA, Medidata Solutions, and IQVIA (Sertkaya et al. 2024)—offers useful benchmarks. Table 10 presents average per patient costs (2025 USD), patient enrollment per study, and the mean number of studies per phase:

- **Phase 1** (adjust by therapeutic area)
    - Cost per patient: $103,500
    - Number of patients per study: 51
    - Mean number of studies: 2

- **Phase 2** (adjust by therapeutic area)
    - Cost per patient: $74,600
    - Number of patients per study: 235
    - Mean number of studies: 2

- **Phase 3** (adjust by therapeutic area)
    - Cost per patient: $67,700
    - Number of patients per study: 630
    - Mean number of studies: 3

We also make cost adjustments based on therapeutic area. Table 10 shows the average cost per study for each phase, calculated using per-patient costs and patient enrollment numbers by specialty (Sertkaya et al. 2024). For detailed per-patient costs and enrollment figures by development phase, see Table 39 in the References section (pages 276-277).

Quick tip: Per-patient costs in clinical trials can vary depending on the therapeutic modality. Small molecules and traditional biologics tend to have greater lower associated costs compared to cell and gene therapies. While there's no publicly available industry benchmark, this trend aligns with differences in manufacturing complexity. For example, cell therapies can cost over $100,000 per patient to manufacture, while gene therapies may range from $500,000 to $1M, excluding R&D expenses (MarketsandMarkets 2023). These higher costs reflect the intricate processes and specialized infrastructure required to develop and produce these advanced treatments.

CLINICAL TRIAL COST PER STUDY BY THERAPEUTIC AREA AND DEVELOPMENT PHASE

Therapeutic Area	Cost Per Study (2025 $M)		
	Phase 1	Phase 2	Phase 3
Anti-infective	1.7	18.4	22.1
Cardiovascular	3.2	10.0	48.7
Central nervous system	4.9	15.2	26.8
Dermatology	4.8	19.2	35.3
Endocrine	4.2	14.8	25.8
Gastrointestinal	3.0	23.7	30.2
Genitourinary system	3.4	18.9	27.2
Hematology	13.8	15.8	35.3
Oncology	7.7	13.8	34.9
Respiratory system	2.8	11.3	30.9
Ophthalmology	7.9	18.5	89.5
Pain and anesthesia	4.2	26.8	93.9
Immuno-modulation	4.5	19.8	21.7
All indications	**5.3**	**17.6**	**42.8**

Source: Sertkaya et al. 2024

Table 10. Analysis of clinical trial costs per study organized by therapeutic area and clinical development phase.

Although less important for your biotech valuation model, the study also reports the total cost per phase in 2025 USD (may include multiple trials per phase), which may provide you with a ballpark comparison for other industry benchmark comparisons:

- **Phase 1:** $9.1M
- **Phase 2:** $26.8M
- **Phase 3:** $113.9M

As with all these studies, there is significant variability in the cost per patient, the number of patients per study, the number of studies conducted, and total drug development cost across different therapeutic areas. The 2024 study has published supplemental analysis, which you may find helpful (Sertkaya et al. 2024). We recommend you adjust your calculations based on the therapeutic area you are planning to enter.

> **Quick tip:** When adjusting the duration and cost of your own trials, it is preferable to use completed comparator trials over making generalized assumptions. Furthermore, early-phase trials are usually more expensive per patient due to challenging recruitment, test complexity, and screening processes, with costs varying significantly by therapeutic area. Note that there can be a high variability of clinical trial costs related to factors such as low prevalence of the patient population, competition, decreased recruitment efficiency, off-label agents purchased from sponsors, low diagnosis rates, and low patient interest.

Drug development success rates advancing a drug from one phase to the next is the cornerstone of drug development and serves as the foundation for our risk-adjusted biotech valuation model. Each phase carries its own probability of success, which can vary significantly.

A 2024 study estimated the likelihood of transitioning between phases as follows (Sertkaya et al. 2024):

- **Phase 1 → Phase 2:** 60%
- **Phase 2 → Phase 3:** 36%
- **Phase 3 → FDA Review:** 66%

Notably, the transition from Phase 2 to Phase 3 has the lowest probability of success, a trend observed across all disease areas. This bottleneck often stems from Phase 2's focus on efficacy testing, where many drugs that showed promise in preclinical animal models fail to demonstrate the desired therapeutic effect in humans. This critical hurdle underscores the inherent uncertainty of drug development and will impact on the valuation of biotech products, especially those that are riskier.

Importantly, these transition percentages include the failure both for scientific reasons and strategic reasons based on budget constraints or business direction.

A 2015 study of 812 oral drug candidates between 2000 and 2010 showed that common strategic (non-scientific) reasons for discontinuing development (Waring et al. 2015):

- **Portfolio Rationalization:** The product no longer aligned with the company's vision, strategic focus, objectives, or business needs—excluding cases where the decision was directly tied to a merger.

- **Commercial Reasons:** Development was halted due to budget limitations, resource constraints, or divestments associated with mergers.

In Table 11, we can see the failure rate of nonscientific reasons for Phase 1 and Phase 2 which is 25% and 24%, respectively (Waring et al. 2015).

RATE OF DRUG DEVELOPMENT TERMINATION DUE TO NONSCIENTIFIC REASONS

Termination Reason	Phase 1	Phase 2
Portfolio Rationalization	18%	21%
Commercial Reasons	7%	3%
Total	**25%**	**24%**

Source: Waring et al. 2015

Table 11. Analysis showing the failure rate breakdown of different nonscientific factors contributing to drug development terminations.

These findings are further supported by another study that showed that in-licensed drugs have a higher probability of success by 16–17% points higher than for self-originated products for Phase 1 and Phase 2 (DiMasi et al. 2016).

If you are presenting your model findings for partnership purposes, consider adjusting your assumptions to reflect that reliable partners are less likely to terminate programs for non-scientific reasons. Excluding these non-scientific failures increases success rates by approximately 20%—for example, boosting Phase 1 to Phase 2 probability of success from 60% to 80% and Phase 2 to Phase 3 from 36% to 56%.

We also sometimes make adjustments based on the therapeutic area—particularly where multiple studies consistently indicate a higher or lower success rate compared to the base case. Table 12 outlines the probability of success for each phase transition by therapeutic area (Sertkaya et al. 2024).

SUCCESS PROBABILITY BY PHASE TRANSITION AND THERAPEUTIC AREA

Therapeutic Area	Transition Success Probabilities			
	Nonclinical to Phase 1*	Phase 1 to 2	Phase 2 to 3	Phase 3 to FDA Review
Anti-infective	68.0%	65.9%	49.6%	74.1%
Cardiovascular	68.0%	65.0%	37.1%	57.6%
Central nervous system	68.0%	61.5%	33.1%	55.9%
Dermatology	68.0%	60.2%	35.9%	65.5%
Endocrine	68.0%	68.0%	46.3%	63.9%

Gastrointestinal	68.0%	71.6%	35.3%	55.3%
Genitourinary system	68.0%	62.9%	44.9%	71.4%
Hematology	68.0%	73.3%	56.6%	75.0%
Oncology	68.0%	61.5%	26.8%	42.7%
Respiratory system	68.0%	68.5%	27.4%	75.6%
Ophthalmology	68.0%	86.0%	52.7%	58.3%
Pain and anesthesia	68.0%	60.2%	35.9%	65.5%
Immuno-modulation	68.0%	69.9%	40.1%	65.4%
All indications	**68.0%**	**60.2%**	**35.9%**	**65.5%**

*Studies analyzing nonclinical to Phase 1 transitions did not stratify data by therapeutic area

Source: Sertkaya et al. 2024

Table 12. Probability of success for each phase transition by therapeutics area.

Additional factors such as drug type, rare versus chronic diseases, and the presence of biomarkers were analyzed. Overall, rare diseases and the use of preselection biomarkers tend to significantly increase success rates, by as much as 10% in certain phases. While prior studies have shown that biologics typically outperform other drug types by over 10% across all phases, this study found that biologics performed more similarly to other drugs, though they still showed a slight advantage over NMEs. Table 13 offers a comparison of these factors compared to all drug indications seen as a percent difference (QLS Advisors 2021).

PHASE TRANSITION SUCCESS PROBABILITY BY DRUG-SPECIFIC FACTORS

Drug-Specific Factors	Transition Success Probabilities (Difference from Baseline)		
	Phase 1 to 2	Phase 2 to 3	Phase 3 to Approval
Baseline (All Drugs)	52.0%	28.9%	57.8%
Biologic	52.5% (+0.5)	32.4% (+3.5)	56.7% (-1.1)
NME	50.6% (-1.4)	25.6% (-3.3)	50.6% (-7.2)
Rare diseases	67.4% (+15.4)	44.6% (+15.7)	60.4% (+2.6)
Chronic diseases	46.0% (-4.0)	23.1% (-5.8)	59.5% (+1.7)
Biomarkers	52.4% (+0.4)	46.3% (+17.4)	68.2% (+10.4)
No Biomarkers	52.0% (+0.0)	28.3% (-0.5)	57.1% (-0.7)

Source: QLS Advisors 2021

Table 13. *Drug-specific factors and their impact on clinical trial success probability relative to baseline by phase transition.*

Table 14 provides a summary of the duration, cost, and probability of success in a condensed chart for your reference (QLS Advisors 2021; Sertkaya et al. 2024).

CLINICAL DEVELOPMENT PHASE:
TIME DURATION, COST, & PROBABILITY OF SUCCESS

Development Phase	Duration	Cost per study (2025 $M)	Probability of Success
Phase 1	2.3 years	$5.3	60%
Phase 2	3.6 years	$17.6	36%
Phase 3	3.3 years	$42.8	66%
Total	**9.2 years** (range 8-11 years)	**$65.8** (range $40-$120)	**14%**

Source: QLS Advisors 2021 and Sertkaya et al. 2024

Table 14. Duration, cost, and probability of success metrics by clinical development phase.

Application Approval (and Phase 4) Stage

After successful clinical trials, biotech companies must seek regulatory approval to bring their drug to market through a final approval stage that determines whether the treatment can be made available to patients, followed by ongoing monitoring studies to ensure the drug's safety and effectiveness in real-world use.

New Drug or Biologics License Application

As we previously discussed in Chapter 1, a NDA seeks approval to market a small molecule drug, while a BLA is for biological products (Allucent 2024). Following successful Phase 3 trials, an NDA or BLA is submitted to the FDA for review. This submission includes comprehensive data from all previous phases to demonstrate that:

- The drug is safe and effective for the proposed use, and the benefits outweigh the risks.

- The labeling is appropriate and contains all necessary information about the drug.

- The manufacturing methods preserve the drug's identity, strength, quality, and purity.

For BLAs, additional emphasis is placed on ensuring the biological product's safety, purity, and potency due to its complexity and derivation from living systems.

The overall application approval success rate is around 88% with an average length of 1.3 years and a range of 1–1.5 years (Sertkaya et al. 2024). Shorter times can be considered for priority or "fast track" applications.

In Table 15, you'll find the success probabilities for application approval by therapeutic areas (Sertkaya et al. 2024). We usually do not make adjustments based on the therapeutic area given the narrow distribution.

SUCCESS PROBABILITY OF APPLICATION APPROVAL BY THERAPEUTIC AREA

Therapeutic Area	Transition Success Probabilities
	FDA Review to Approval
Anti-infective	94.4%
Cardiovascular	75.5%
Central nervous system	87%
Dermatology	88.3%
Endocrine	81.3%
Gastrointestinal	86.2%

Genitourinary system	85.7%
Hematology	84%
Oncology	85.5%
Respiratory system	89.5%
Ophthalmology	77.5%
Pain and anesthesia	88.3%
Immuno-modulation	95.3%
All indications	**88.3%**

Source: Sertkaya et al. 2024

Table 15.   Success probability of FDA review to approval by therapeutic area.

For drugs that require a full application (e.g., new active ingredient, indication, non-orphan status, etc.), costs can run up to $4.5M (2025 USD) per application, which includes filing fees by regulatory agencies, expenses for outside consultants or employees to prepare applications, etc. (Beaney 2024). In contrast, costs in the European Union are estimated to be up to $1M per application (2025 USD) (European Medicines Agency 2025; DLRC 2024), and costs in Japan are even lower at around $250,000 per application (2025 USD) (PharmaInfo 2023). Application costs in general have increased since post-COVID-19. In addition, most regions require an annual regulatory filing maintenance fee, but these are largely ignored in forecasts during the commercialization period given the small amount, which is less than $1M in total (Federal Register 2024).

Phase 4 (Post-Market Surveillance)

Phase 4, also known as post-market surveillance, takes place after regulatory approval and market launch. Its purpose is to collect additional data on a drug's risks, benefits, and optimal use in real-world settings, particularly across larger and more diverse populations.

For example, new adverse events or rare side effects can emerge at this stage—effects that may not have been detected in smaller scaled studies during clinical trials. A well-known example is Merck's withdrawal of its blockbuster arthritis drug, Vioxx (rofecoxib), in 2004 after Phase 4 studies revealed an increased risk of heart attacks and strokes—risks not fully evident in its large Phase 3 trial (Singh 2004).

Phase 4 studies show the following characteristics:

- ⊿ **Phase 4** (adjust by therapeutic area)
  - ✦ Cost per patient: $44,800
  - ✦ Number of patients per study: 708
  - ✦ Mean number of studies: 2

Phase 4 trial costs average approximately $31.8M per trial (2025 USD), with a mean of two trials typically conducted over a drug's lifetime (Sertkaya et al. 2024). However, costs per trial can vary significantly, ranging from $8M to $90M. Given this variability, your Phase 4 cost estimate should depend on the availability of a suitable comparator trial based on the following conditions:

1. **If a suitable comparator Phase 4 trial is identified based on indication,** base your assumptions on its number of patients and duration, using Phase 3 per-patient trial costs to estimate the total expense.

2. **If no suitable comparator is available,** assume Phase 4 trial expenses will be similar to Phase 3 trial expenses or use comparable industry benchmark data from Sertkaya et al. (2024), as seen in Table 16. For detailed per-patient costs and enrollment figures for Phase 4 trials, see Table 39 in the References section (pages 276-277).

3. **If no significant Phase 4 trials are expected,** a constant annual R&D expenditure of $5M (range of $2M to $12M) during the market exclusivity period can be applied. Assuming $5M in annual post-approval R&D expenses over a 14-year market exclusivity period totals approximately $70M. This amount roughly equals the cost of conducting two Phase 4 trials at $35M each, aligning with typical industry practice of one to two Phase 4 trials per drug and consistent with industry cost estimations (Sertkaya and Franz 2022). This model assumes the following constant R&D spend:

a. The company has completed major clinical trials and received regulatory approval.

b. No significant new indications or formulations are being actively pursued.

c. Ongoing activities focus on regulatory compliance and small post-marketing studies for additional regulatory submission.

PHASE 4 CLINICAL TRIAL COST PER STUDY BY THERAPEUTIC AREA

	Cost Per Study (2025 $M)
Therapeutic Area	Phase 4
Anti-infective	25.3
Cardiovascular	22.0
Central nervous system	16.0
Dermatology	36.0
Endocrine	35.0
Gastrointestinal	90.6
Genitourinary system	8.8
Hematology	22.1
Oncology	7.9
Respiratory system	28.1
Ophthalmology	12.7
Pain and anesthesia	14.9
Immuno-modulation	14.8
All indications	**31.8**

Source: Sertkaya et al. 2024

Table 16. *Analysis of Phase 4 clinical trial costs per study organized by therapeutic area.*

Table 17 provides a summary of the duration, cost, and probability of success in a condensed chart for your reference (QLS Advisors 2021; Beaney 2024; Sertkaya et al. 2024).

APPROVAL AND PHASE 4 POST-MARKET SURVEILLANCE TRIAL: TIME DURATION, COST, & PROBABILITY OF SUCCESS

Development Phase	Duration	Cost (2025 $M)	Probability of Success
Approval	1.3 years (range 1-1.5 years)	$4.5	88%
Phase 4 or Post-market Surveillance	Remainder of product life	Comparator Phase 4 trial size and duration + Phase 3 trial per-patient costs Or Similar to Phase 3 trial expenses Or Use industry benchmark: 31.8 per Phase 4 trial (range $8-$90) Or $5 per year during market-exclusivity period if no Phase 4 trial planned (range $2-$12)	-

Source: QLS Advisors 2021, Beaney 2024, and Sertkaya et al. 2024

Table 17. *Summary of regulatory application approval and Phase 4 post-marketing study duration, cost, and success rates.*

Applying Probability of Success in Valuation

The probability of success is necessary for risk-adjusting our valuation model because it helps us account for the likelihood that a drug or treatment will successfully progress through each stage of development, from clinical trials to regulatory approval. In our model, we define these terms as follows:

- **Probability of occurrence**: The likelihood that a biotech product will progress based on its current phase of development.

- **Probability of success**: The expected chance of successful phase progression, derived from industry benchmarks.

- **Cumulative probability of success**: The overall success rate for a given phase, calculated as the product of the probability of occurrence and the probability of success.

Let's walk through a simple example to show how these probabilities work in practice—by multiplying the success rates of each remaining clinical phase step by step. This calculation approach will help you build more accurate financial models that are properly risk-adjusted, accounting for the real-world uncertainties of drug development.

Example Step-by-Step Calculation

We begin by identifying the biotech product's current clinical development phase and then evaluate its likelihood of progressing through the remaining stages. Each phase has a specific probability of success, informed by historical industry benchmarks, which in turn shape our estimates for probability of occurrence and cumulative probability of success.

To determine the cumulative probability of success, we multiply the probability of occurrence (the likelihood of progressing to the next phase) by the probability of success (the industry benchmark for successfully completing that phase). This method assumes that each phase is an independent hurdle the drug must clear.

Suppose a drug has completed Phase 1 and is now advancing to Phase 2:

- ⅄ **Probability of Occurrence for Phase 2**: 100% (since the drug has already passed Phase 1)
- ⅄ **Probability of Success for Phase 2**: 36% (based on industry data)
- ⅄ **Cumulative Probability of Success for Phase 2**: 1.0×0.36 = 0.36 or **36%**

Now, if the drug successfully progresses to Phase 3:

- ⅄ **Probability of Occurrence for Phase 3**: 36% (if Phase 2 is passed)
- ⅄ **Probability of Success for Phase 3**: 66% (based on industry data)
- ⅄ **Cumulative Probability of Success for Phase 3**: 0.36×0.66 = 0.24 or **24%**

Finally, if the drug advances to FDA approval:

- ⅄ **Probability of Occurrence for Approval**: 24% (if Phase 3 is passed)
- ⅄ **Probability of Success for Approval**: 88% (based on industry data)
- ⅄ **Cumulative Probability of Success for Approval**: 0.24×0.88 = 0.21 or **21%**

By calculating the probabilities, you'll be able to accurately track the probability of success at each development stage for your specific assets. Armed with these probability figures, you can build valuation models that properly discount future cash flows based on real risk factors rather than guesswork. This approach will help you make more confident investment decisions, set appropriate milestone valuations for partnerships, and communicate more effectively with investors who expect sophisticated risk analysis. We'll explore how to incorporate these probabilities into your complete risk-adjusted valuation model in Chapter 6.

CASE STUDY
Rare Neurological Disease Biologic

To ground these figures in a real-world context, let's consider a scenario you might encounter as a biotech investor or founder.

Imagine you're evaluating Company B's lead asset NBX-472, a monoclonal antibody targeting a rare neurological disorder that has just completed Phase 1 with promising safety data. The company is now preparing for Phase 2, and you need to determine a reasonable valuation for a potential investment or partnership.

Looking at our drug-specific factors, NBX-472 has several positive attributes:

- It's targeting a rare disease (shown in the table with higher success rates)

- It's a biologic (slightly better than baseline in Phase 2 to 3)

- The Phase 2 trial will incorporate a biomarker for patient selection

For phase transitions where multiple drug-specific success factors are relevant, we apply only the factor showing the greatest difference from baseline.

Adjusting our base probability estimates for central nervous system indications using the applicable factors from Table 13:

- **Base Phase 1 → Phase 2 success:** 100% (already achieved)

- **Base Phase 2 → Phase 3 success:** 33.1% + 17.4% (biomarker factor is largest) = 50.5%

- **Base Phase 3 → FDA Review success:** 55.9% + 2.6% (rare disease factor is largest) = 58.5%

- **FDA Review → Approval success:** 88%

Cumulative probability of success: 50.5% × 58.5% × 88.0% = 26.0%

This is approximately twice the industry average of 12.5% from Phase 1 to approval.

The development timeline and costs for neurology/central nervous system, obtained from Figure 31 (phase durations), Table 10 (costs per study), and Table 17 (application costs), are presented below:

- **Expected Phase 2 timeline:** 3.7 years

- **Expected Phase 2 cost:** $15.1M

- **Expected Phase 3 timeline:** 3.7 years

- **Expected Phase 3 cost:** $26.8M

- **Application to Approval timeline:** 1.6 years

- **Application cost:** $4.5M

Total timeline until approval: 3.7 + 3.7 + 1.6 = 9.0 years

Total cost until approval: $15.1M + $26.8M + $4.5M = $46.4M

Let's calculate the risk-adjusted cost by multiplying each phase's cost by its probability of occurrence and success:

- **Phase 2 cost** (100% probability of occurrence × 50.5% probability of success = 50.5%): $15.1M × 0.505 = $7.63M

- **Phase 3 cost** (50.5% probability of occurrence × 58.5% probability of success = 29.5%): $26.8M × 0.295 = $7.91M

- **Application cost** (29.5% probability of occurrence × 88% probability of success = 26.0%): $4.5M × 0.260 = $1.17M

Total risk-adjusted cost: $7.63M + $7.91M + $1.17M = $16.7M

When building a risk-adjusted valuation model for NBX-472, the R&D components can help map out the foundational blocks of your valuation:

1. The 26.0% overall probability of success gives you a useful to value for risk adjustment of future revenues during commercialization.

2. The 9-year development timeline maps out when to expect costs and when revenues might be expected to begin.

3. The $46.4M in nominal development costs shows your total potential investment needs.

4. The $16.7M risk-adjusted development cost provides a risk-adjusted view of expected R&D expenditure.

These numbers demonstrate why rare disease assets with biomarkers often command premium valuations despite much smaller market sizes. The combination of rare disease targeting, biomarker use, and Phase 1 completion creates a meaningful improvement in the cumulative probability of success compared to the industry average, potentially more than doubling the chances of reaching approval. This could significantly impact the expected returns despite the lengthy development timeline typical for neurological indications.

However, these calculations represent risk-adjusted probabilities but don't account for the time value of money—the principle that a dollar today is worth more than a dollar received years from now. For instance, if NBX-472 is projected to generate $100M in annual peak sales after approval, with a 26% overall probability of success, the risk-adjusted expected revenue would be $26M annually. However, this figure only reflects development risk, not timing.

In practice, you must consider both the risk-adjusted revenue and the substantial time delay before those revenues materialize. NBX-472 faces a 9-year development timeline with most costs incurred upfront, while peak sales won't begin until after approval. This timing mismatch significantly impacts valuation. The next chapter will explore how to apply appropriate discount rates to these risk-adjusted figures for accurate valuation.

What to Keep in Mind

Bringing a drug from discovery to market is quite the journey—complex, costly, and filled with both challenges and opportunities. Throughout this chapter, we've explored what really drives success in biotech R&D: therapeutic indication choice, smart use of biomarkers, and selecting the right drug modality, among other factors.

That 13% probability of progressing from Phase 1 to approval? It's a reality check we all face in this industry. But rather than seeing it as discouraging, think of it as your compass for making smarter plans and decisions as you navigate the development process.

You're now equipped with practical knowledge about timelines, costs, and risks at each clinical phase—insights that transform how you'll evaluate the business case for any biotech project. This understanding opens the door to risk-adjusted valuation, helping you forecast returns in a way that acknowledges both the potential rewards and the very real hurdles along the way.

In the next chapter, we'll explore the remaining financial components of your valuation toolkit: operating expenses, discount rates, and payment structures (including royalties, upfront payments, and milestone payments).

BEFORE YOU WRAP UP

Take a moment to reflect on what you've learned about research and development (R&D) timelines, costs, and probabilities of success. These key takeaways and questions are designed to help you connect these concepts to your biotech valuation model and identify areas you'd like to explore further.

Key Takeaways

- Success in biotech R&D is far from random—specific factors such as therapeutic area, biomarker use, and drug modality can significantly improve your odds of reaching approval.

- The industry average 13% success rate from Phase 1 to approval conceals enormous variation—from 24% in hematology down to 5% in oncology and urology—making therapeutic area selection a critical business decision.

- Phase 2 represents the valley of death in drug development, with only 36% of candidates advancing to Phase 3, primarily due to efficacy failures that weren't predicted by animal models.

- Development timelines and costs vary dramatically by therapeutic area, which will be essential for accurate valuation and forecasting.

- Regulatory factors can meaningfully impact clinical development timelines—accelerated approval can shorten timelines by 3 years, while orphan designation may extend them by 1.5 years despite smaller trial sizes.

- Non-scientific reasons account for approximately 25% of development terminations, suggesting that in-licensing candidates may offer significantly improved success probabilities.

BEFORE YOU WRAP UP *(Continued)*

Reflection Questions

1. Which success factors in your drug candidate's profile might enhance its probability of approval compared to industry averages? Consider therapeutic area, modality, biomarker strategy, and regulatory pathway.

2. How would your valuation change if you adjusted your probability of success to reflect only scientific (not strategic) termination reasons?

3. What is your product's cumulative probability of success from its current stage to approval, and how does this compare to industry averages?

4. How might your development timeline be compressed through strategic use of FDA programs like breakthrough therapy designation or accelerated approval?

5. If you're developing a product in an area with lower success rates (like neurology or oncology), what specific strategies could you implement to improve probability of success?

6. How would doubling either the timeline or cost of your Phase 3 trial impact your overall valuation? Which has a greater effect?

05

Operating Expenses, Discount Rates, and Payment Structures

A s with any business, financial sustainability and market impact hinge on a company's ability to manage expenses, optimize revenue streams, and assess long-term returns effectively. Biotech companies, however, face unique operational challenges due to lengthy development timelines, rigorous regulatory requirements, and complex partnership ecosystems.

Operating expenses, discount rates, and payment structures are collectively essential to biotech valuation because they bridge the gap between current operations and future value creation. Each element serves a distinct but interconnected purpose: operating expenses reveal the efficiency of your commercial infrastructure and organizational capabilities; discount rates capture the unique risks and extended timelines inherent in biotech investments; and payment structures in licensing deals and partnerships determine how value flows between industry participants. Together, these components provide the foundation for assessing market readiness, securing strategic partnerships, and managing the complex risk profile that defines biotech investment success.

This chapter will delve into these key financial components, providing practical frameworks that show how these elements shape investment decisions, partnership negotiations, and ultimately, company valuations in the biotech sector.

Breakdown of Selling, General, and Administrative (SG&A) Expenses

SG&A expenses typically encompass two main components:

- **Selling**: Costs associated with commercialization efforts such as sales and marketing.

- **General and Administrative (G&A)**: Costs that cover infrastructure and overhead costs essential for the company's operations.

Each of these components can encompass various types of expenses as seen in Table 18, which outline common examples:

BREAKDOWN OF SG&A EXPENSE COMPONENTS

Selling	General & Administrative
Direct-to-consumer Sales	Executive Compensation
Direct-to-physician Sales	Legal Fees
Promotions	Corporate Headquarters
Samples to Physicians	Employee Salaries
Educational and Promotional Meetings	Human Resources (HR)
Advertisements	Information Technology (IT)
Literature Publications	

Table 18. Breakdown of SG&A components showing sales and general and administrative categories with examples.

As with any costs, SG&A expenses can significantly impact a company's profitability, as illustrated in Figure 32 showing the profit and loss of the top 150 largest publicly listed global biotech companies (Baur et al. 2023). You may recognize terms such as Cost of Goods Sold (COGS) and Research & Development (R&D), which we have discussed in previous chapters. Another important metric to consider is operating income, also referred to as operating profit or earnings. This financial measure reflects the profit generated from a company's core business operations, excluding any income derived from non-operational activities.

GLOBAL BIOTECH INDUSTRY PROFIT AND LOSS STRUCTURE, 2008 TO 2022

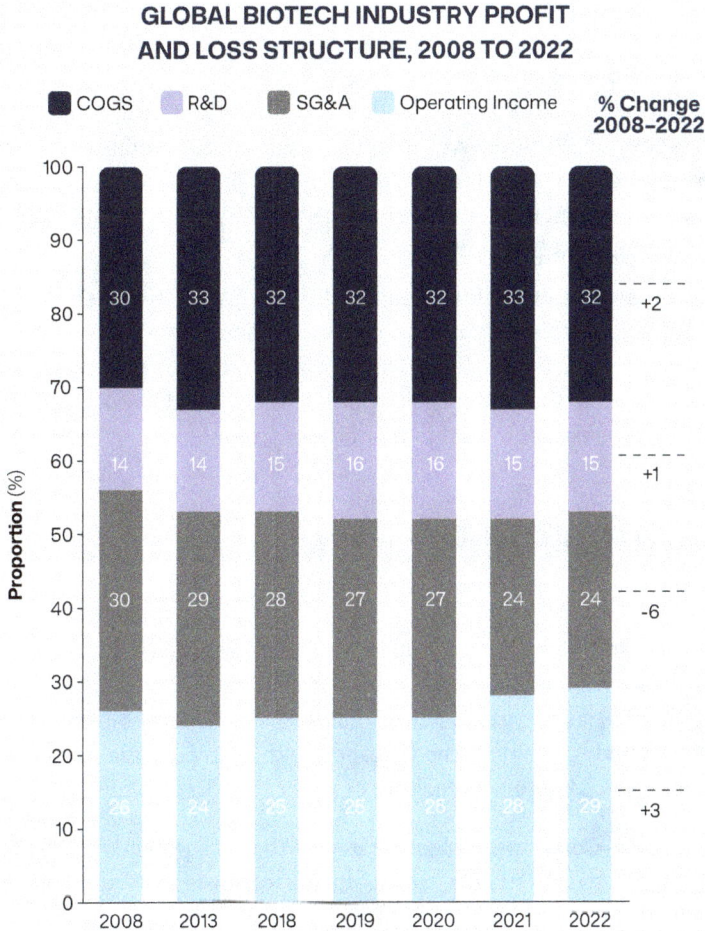

Source: Baur et al. 2023

Figure 32. Profit and loss structure of the global biotech industry from 2008 to 2022.

The data by Roland Berger indicates that recent years have been challenging in the post-COVID-19 era. Financial pressures have intensified, and biotech companies have focused on optimizing operations by reducing SG&A expenses, primarily in sales and marketing. Notably, the analysis showed no significant difference in the share of revenue allocated to SG&A based on company size. SG&A accounts for approximately 24% of revenue for a large-cap and mature biotech company.

In the following sections, we will outline a method for forecasting SG&A expenses and its individual components (i.e., sales, marketing and G&A). This big picture approach is structured as follows:

- **Pre-Launch Period:** We forecast each SG&A component separately using industry benchmarks, accounting for the distinct cost drivers and investment patterns typical of early-stage companies across sales, marketing, and G&A activities.

- **Commercial Period:** From launch through Year 3, we forecast each SG&A component separately using industry benchmarks while revenue remains below a defined threshold. Once revenue reaches this threshold at any point during the commercial period, we transition to calculating total SG&A expenses as a fixed percentage of revenue

Estimating Sales and Marketing Costs

As previously stated, sales and marketing are necessary for the successful commercialization of any biotech product. Otherwise, how will the public know that the company has released a new product? These expenses encompass various activities designed to promote the therapeutic to both consumers and healthcare professionals. Key components include:

- **Direct-to-Consumer Sales**: Strategies to reach potential patients directly through advertisements, social media campaigns, and public relations efforts.

- **Direct-to-Physician Sales**: Activities aimed at engaging physicians through face-to-face interactions, educational materials, and samples.

- **Promotions**: Initiatives such as discounts, loyalty programs, and co-pay assistance programs to encourage product adoption.

- ⋏ **Samples to Physicians**: Providing free samples to doctors so they can try the therapeutic and recommend it to their patients.
- ⋏ **Educational and Promotional Meetings**: Organizing seminars, workshops, and conferences to educate healthcare professionals about the therapeutic benefits and usage.
- ⋏ **Advertisements**: Utilizing various media channels (print, online, TV, radio) to raise awareness about the therapeutic.
- ⋏ **Literature Publications**: Publishing articles and papers in medical journals and other platforms to highlight the therapeutic efficacy and safety.

Estimating Sales Expenses

The primary expense that we will include in the valuation model is estimated by using the sales force, which is the team responsible for driving sales and the vital link between industry and healthcare providers (Chaganti 2023). This allows us to be transparent about the overall commercial costs, easily adjust depending on partnerships and commercialization models, and promote synergies when a biotech company is acquired or uses existing sales reps from an existing partner.

For early-stage biotech companies, estimating commercial costs via calculating the sales force is more-than-adequate given commercialization is far in the future as well as early-stage investors are not interested in this level of detail but rather the methodology. Accordingly, we will exclude considerations of commercial team structures, physician numbers, geographic distribution, and the presence of trained medical science liaisons who support sales representatives through educational efforts.

Our sales expense estimation approach follows four simple steps:

1. Size the sales force based on therapeutic area and physician specialties.
2. Apply standard cost rates per sales representative.
3. Distribute expenses across the product lifecycle, including pre-launch preparation.
4. Adapt U.S.-based figures for international markets as needed.

Cost

For most biotech companies, the cost of the sales force is calculated based on the estimated annual number of sales representatives required and the annual cost per representative. These costs vary depending on whether the therapeutic targets a primary care, specialty, or hospital setting, and they are projected over the entire commercialization period.

In Table 19, we provide the latest sales force cost per person by specialty area and setting, which includes bonus, payroll fee, and additional benefits (Glassdoor 2025; Dunleavy 2025; Becker 2023; Thompson 2024):

SALES FORCE EXPENSES:
PRIMARY CARE VS. SPECIALTY/HOSPITAL SETTINGS

Sales Rep	Primary Care	Specialty/Hospital
Total Number	300-600	50-200
Cost for Each* (2025 $)	$150,000-200,000	$300,000-350,000

* Includes base salary, benefits, insurance coverage, retirement plans, and other financial incentives

Source: Glassdoor 2025, Dunleavy 2025, Becker 2023, and Thompson 2024

Table 19.   Comparison of sales force size and expenses between primary care and specialty/hospital market settings.

To estimate sales expenses for valuation purposes, we must consider the complete lifecycle of sales force deployment and associated expenses. This timeline-based approach captures three critical phases that impact overall sales cost:

Pre-launch Period

Sales force investment strategically begins approximately one year before product launch. During this preparatory phase, biotech companies allocate SG&A budget to build commercial infrastructure and initiatives to establish market presence.

Our financial models typically account for 50% of the eventual sales team during this period, focusing on hiring key commercial leadership, training the initial sales cohort, and developing territory management strategies. These pre-launch investments, while potentially significant, are essential for ensuring rapid market uptake once approval is secured.

Commercial Period: Market Exclusivity

Upon product launch, the sales force reaches full capacity and remains at 100% staffing throughout the commercial exclusivity period. This represents the most substantial and sustained sales expense in the product lifecycle. During this phase, the total annual sales force expense is calculated by multiplying the number of representatives by the fully loaded cost per representative (as detailed in Table 19). This approach provides a straightforward estimation of one of the largest components of SG&A expenditure.

Commercial Period: LOE

Following LOE, sales force expenses are typically eliminated from valuation models as generic competition significantly reduces both market share and profit margins. While some companies may choose to maintain limited sales support for branded products despite generic entry, such strategies rarely generate sufficient returns to justify inclusion in early-stage valuation models. For most biotech valuation purposes, we assume complete elimination of the sales force upon LOE.

Growth of Expenses

Compensation growth follows national compensation trends, maintaining an approximate 2% annual increase.

> **Quick tip:** We do not scale the sales force by the number of target physicians because that number can vary by thousands across specialties. Instead, we focus on high-priority targets such as physicians who are heavy prescribers or key influencers who are accessible to sales reps. These individuals represent only a small fraction of the total physician population but have a disproportionate impact on prescribing behavior.

Estimating Marketing Expenses

Unlike sales force calculations, marketing expenditures require a different approach due to limited public disclosure of product-specific investments. Let's walk through how to build a practical marketing expense model using industry benchmarks.

Cost

A 2021 study broke down the revenue of public U.S. manufacturers by its components (Jiang et al. 2022). As seen in Figure 33, the data showed advertising expenses, defined as costs of producing and distributing advertisements and sponsoring public events to promote a brand or product, remaining relatively constant around 3% of revenue across the pharmaceutical industry over the past decade (Jiang et al. 2022). Given that this is pre-COVID data of public biotech companies and may not capture all marketing cost components, actual expenditure is likely higher for growth-stage companies that have not yet commercialized their products.

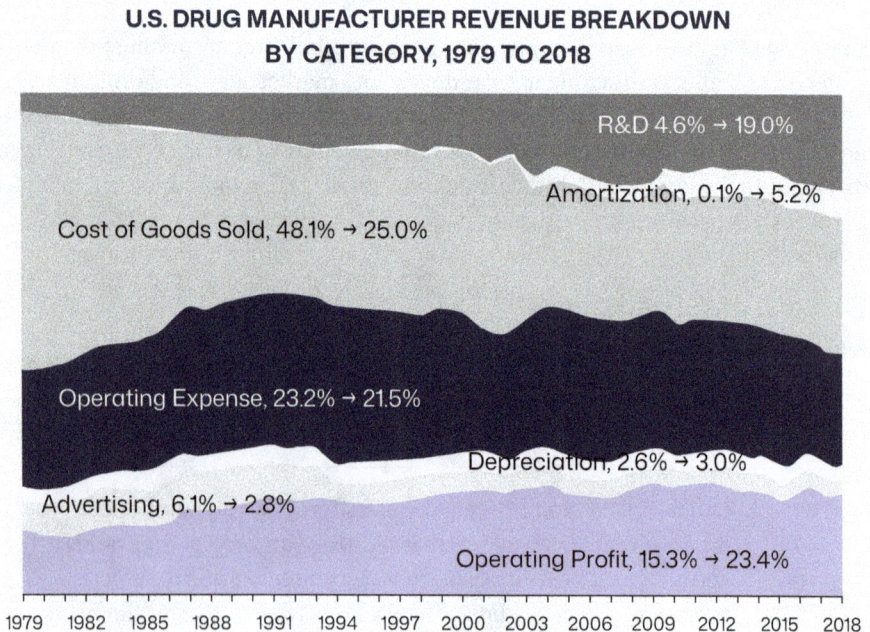

U.S. DRUG MANUFACTURER REVENUE BREAKDOWN BY CATEGORY, 1979 TO 2018

R&D 4.6% → 19.0%

Amortization, 0.1% → 5.2%

Cost of Goods Sold, 48.1% → 25.0%

Operating Expense, 23.2% → 21.5%

Depreciation, 2.6% → 3.0%

Advertising, 6.1% → 2.8%

Operating Profit, 15.3% → 23.4%

1979 1982 1985 1988 1991 1994 1997 2000 2003 2006 2009 2012 2015 2018

Source: Jiang et al. 2022

Figure 33. Revenue breakdown analysis for publicly traded U.S. drug manufacturers from 1979 to 2018.

An interesting trend to note is that while marketing spending has remained stable, R&D investment has grown dramatically from 4.6% of revenue in 1979 to 19.0% by 2018, reflecting the industry's expanding push for innovation through both internal research and strategic acquisitions (Jiang et al. 2022).

However, calculating marketing as a percentage of revenue doesn't accurately reflect the reality of drug commercialization. Marketing costs are typically highest during prelaunch and early launch phases when revenue is still minimal, while high-revenue products often experience diminishing returns from proportional marketing spend increases.

For our model, we recommend calculating marketing expense as a percentage of sales force costs rather than revenue. Based on vendor contracts, industry benchmarks, and sales force sizing studies, marketing expense typically ranges from 30–70% of sales force expense (Locust Walk Institute 2017). This covers market research, medical affairs, direct-to-consumer and direct-to-physician marketing, digital campaigns, patient assistance programs, key opinion leader engagements, physician education programs, and promotional meetings.

To accurately project marketing expenses for valuation purposes, we must examine how these investments evolve throughout a product's commercialization, similar to sales expenses. This approach identifies three distinct phases where marketing strategies and associated costs change:

Pre-launch Period

Marketing activities begin approximately one year before product launch and are calculated as 30–70% of sales force costs. During this preparation phase, teams focus on building outreach programs, developing marketing materials, and scheduling physician education meetings. The purpose is to support rapid market uptake once regulatory approval is secured.

Commercial Period: Market Exclusivity

Marketing runs at full capacity from product launch through LOE. During this period, annual marketing costs are calculated as 30–70% of sales force costs. This represents the most substantial and sustained marketing expense in the product lifecycle.

Commercial Period: LOE

For early-stage valuations, we recommend eliminating marketing expenses after LOE. While some companies continue promoting branded products and blockbusters drugs, this strategy does not necessarily generalize to all therapeutics, and it is not as applicable during the early stage (Kornfield et al. 2013).

Other Factors to Consider

When estimating marketing expenses, several factors should be considered:

- **High Competitive Intensity:** Markets crowded with similar products often require aggressive marketing efforts to stand out.

- **Established Legacy Competitors:** Facing entrenched brands, especially inexpensive or off-patent options, can demand greater investment to compete effectively.

- **Low Awareness of the Disease or Drug Category:** Educating both consumers and providers about new treatments or conditions may necessitate a higher marketing spend.

- **Limited Innovation in the Clinical Area:** Differentiating a product in a field with minimal breakthroughs often requires additional effort.

- **Diverse and Widespread Provider Base:** Large, heterogeneous groups, such as primary care physicians, often require broader and more tailored outreach.

In these situations, a higher projected marketing budget of 50–100% of sales force costs may be appropriate to account for increased direct-to-consumer advertising, direct-to-physician outreach, and digital marketing requirements.

Estimating General and Administrative (G&A) Costs

G&A expenses are essential to support the infrastructure, management, and operational needs of a biotech company. Key components include:

Executive Compensation

Salaries, bonuses, and stock options for company leadership, including CEOs, CFOs, and other executives.

Legal Fees

Costs associated with regulatory compliance, intellectual property protection, contract negotiations, and other legal needs.

Corporate Headquarters

Rent, utilities, and maintenance for office spaces, as well as associated administrative costs.

Employee Salaries

Wages for non-sales and non-R&D staff, including finance, operations, and administrative roles.

Human Resources

Expenses related to recruitment, onboarding, employee training, and performance management.

Information Technology

Costs for maintaining essential software, hardware, cybersecurity, and data management systems.

Estimating the sales and marketing expenses for an individual drug is relatively straightforward. However, calculating G&A expenses alone can be challenging due to their less clearly defined components. In addition, most companies report only SG&A, making it even more difficult to make comparisons via industry benchmarks. Therefore, we recommend using the following method to estimate SG&A by splitting into two different periods to prioritize benchmark data.

Pre-Launch and Early Commercial Period (Launch through Year 3)

Understanding G&A expenses during the early phases of biotech product development requires examining three key research studies that provide essential benchmarks for financial modeling:

1. A McKinsey & Company study of early-stage biotech companies reveals that G&A expenses consistently represent a significant portion of total expenses (Bleys et al. 2022). For Preclinical Stage biotechs, G&A expenses range from 25–35% of total expenses, with a median of 30%. Clinical-stage companies show slightly higher ratios, ranging from 25–40% with a median of 32%. This consistency across development stages provides a reliable foundation for modeling pre-launch expenses.

2. Pharmagellan's analysis of small- to medium-cap biotech companies introduces an important revenue threshold that affects expense calculations (David et al. 2017). When companies achieve revenues of $\geq$ $540M, their SG&A expenses stabilize at 30% of revenue. Indeed, the SG&A of biotech companies have been noted to be around 30% (Lee 2021; Damodaran 2025). However, companies below this threshold exhibit much greater variability, with SG&A ranging from 10–120% of revenue (David et al. 2017; Chang et al. 2020). This wide variance likely reflects the differing operational strategies of sales and marketing expense in the early launch and low-revenue periods.

3. The Health Advances study identifies Year 3 post-launch as a critical inflection point for biotech companies (Davitian et al. 2018). By this time, companies have sufficient market data to evaluate whether their products are underperforming, meeting expectations, or exceeding projections. This performance assessment drives strategic cost structure adjustments, such as companies less willing to maintain high SG&A spending on underperforming products.

> **MODELING APPROACH**
> For the pre-launch period and the first 3 years post-launch, we recommend using 20–30% of total expenses for G&A calculations. However, if unadjusted revenue reaches ≥ $540M during Years 1–3, switch to calculating SG&A as 30% of revenue rather than calculating G&A directly.

Established Commercial Period (Year 4+ and LOE)

Once a product enters its established commercial phase—typically Year 4 post-launch and continuing through LOE—the financial dynamics shift toward more predictable revenue-based expense structures. During this mature phase, companies have resolved most operational scaling issues and established clear market positioning.

> **MODELING APPROACH**
> Calculate SG&A as 30% of revenue throughout this period, rather than calculating G&A expenses directly. This approach reflects the more stable operational environment and clearer cost-revenue relationships that characterize established commercial products.

Summary

Modeling SG&A Expenses at Different Stages

Now that we've covered how to estimate sales, marketing, and G&A expenses, we have all the necessary components to model SG&A from pre-launch to market. Table 20 summarizes how you would estimate SG&A expenses (and its components) based on where your product is in its lifecycle. To get started:

1. Identify the period your product is in, such as pre-launch (no revenue), early commercial (based on unadjusted revenue levels), or established commercial.

2. Use the rows to guide what costs to include, you'll see how to estimate G&A, sales, marketing, and SG&A at each point.

3. Ensure to adjust your model as the product accordingly throughout its lifecycle. For example, costs will ramp up as you approach launch, then taper off after exclusivity ends.

This approach keeps your SG&A estimates grounded in industry benchmarks but flexible enough to adapt to your specific product.

SG&A EXPENSE CALCULATION FRAMEWORK
BY PRODUCT LIFECYCLE STAGE

Stage of Product Lifecycle	G&A	Sales	Marketing	SG&A
Pre-Launch *No revenue*	20-30% of total expenses, covers core operations before launch.	50% of fully loaded sales force cost, applies only in the year before launch.	30-70% of sales force cost, applies only in the year before launch.	Add G&A + Sales expenses, no separate marketing cost at this stage.
Early Commercial (Years 1-3) *Market Exclusivity*				
Scenario #1 (Unadjusted Revenue < $540M, 2025 $)	20-30% of total expenses.	Equal to fully loaded sales force cost.	30-70% of sales force cost.	Add G&A + Sales + Marketing expenses.
Scenario #2 (Unadjusted Revenue ≥ $540M, 2025 $)	-	-	-	30% of revenue.
Established Commercial *Market Exclusivity to LOE*	-	-	-	30% of revenue.

Table 20. Summary of calculation methods for G&A, sales, marketing, and SG&A expenses at different stages from pre-launch to market through LOE.

Other Company-Specific Financial Terms

In addition to SG&A, several other company-specific financial terms play a key role in biotech valuation—especially when assessing overall company health and future value. Let's take a closer look at three of the most important ones (Wall Street Prep 2024a, 2024b, 2024c):

1. **Working Capital** represents the funds a company needs to run its daily operations. It's calculated as:

    ## Working Capital = Current Assets – Current Liabilities

    - **Current Assets** may include:
        - **Cash:** Money on hand, including investments and short-term assets.
        - **Accounts Receivable:** Money owed by customers for products or services provided.
        - **Inventory:** Unsold goods, including raw materials and finished products.

    - **Current Liabilities** include:
        - **Accounts Payable:** Outstanding invoices for supplies, utilities, taxes, etc.
        - **Short-Term Debt:** Loans or credit due within the year.

    Positive working capital means the biotech company has more short-term assets (like cash or receivables) than short-term liabilities (like bills or loans). This indicates the company can cover its immediate expenses, fund ongoing research, and continue operations smoothly.

    Negative working capital means the company owes more than it owns in the short term. This could signal potential cash flow issues, making it harder to pay bills or fund research, although some biotech companies can manage this if they have strong future funding or partnerships.

2. **Capital Expenditure (CapEx)** represents investments in long-term assets necessary for a company's growth and operations, such as buildings, equipment, or machinery. Examples include:

⅄ **R&D Equipment:** Tools and technology needed for developing new products.

⅄ **Facilities:** Investments in lab or office spaces required for operations.

Higher CapEx means the company is investing more in long-term assets like advanced research equipment or expanding facilities. This can indicate a focus on growth, new product development, or scaling operations but may also lead to higher upfront costs.

Lower CapEx means the company is spending less on long-term assets, which might suggest it is in a more stable phase or prioritizing operational efficiency over expansion. However, too little CapEx could limit the company's ability to innovate or grow in the future.

3. **Depreciation and Amortization (D&A)** help businesses recognize the cost of long-term assets over time, ensuring that expenses are matched with the revenue the assets help generate.

⅄ **Depreciation:** Allocates the cost of physical assets (e.g., machinery, equipment) over their useful life, as these assets lose value over time.

⅄ **Amortization:** Works similarly to depreciation but applies to intangible assets (e.g., patents, trademarks). The cost is spread out over the asset's useful life.

Both processes ensure that companies accurately match the expenses of using assets with the revenue those assets generate.

These financial variables will feed directly into the SOTP valuation model we'll build in the next chapter, helping you better assess the full value of a biotech company.

How Discount Rates Shape Investment Decisions

The discount rate—sometimes called the hurdle rate—represents the minimum return an investor expects for committing capital to a project. In other words, the discount rate usually should reflect the cost of capital, or more specifically, the Weighted Average Cost of Capital (WACC), the defined rate a company must pay its debt holder and equity investors. WACC captures both non-R&D financial risks—such as financing constraints, partnership uncertainty, manufacturing hurdles, and IP concerns—and the inherent R&D risks associated with drug development.

However, in biotech valuation, R&D risk is typically accounted for separately by adjusting future cash flows based on the probability of success at each stage of drug development. Therefore, using WACC as the discount rate would result in double-counting risk in a risk-adjusted model, leading to an inappropriately high discount rate estimate. In practice, the discount rate used in risk-adjusted models should reflect non-drug development risks only, such as financing challenges, partnership issues, manufacturing hurdles, intellectual property concerns, competitive pressures, pricing constraints, reimbursement difficulties, and market adoption barriers. Based on investor surveys and industry benchmarks, we recommend applying stage-specific discount rates that align with the asset's development phase (Rottgen 2025; Stasior et al. 2018):

- **Discovery or Preclinical Stage:** 19-28%
- **Phase 1:** 18-22%
- **Phase 2:** 16-20%
- **Phase 3:** 13-17%
- **Registration:** 10-15%
- **Approval:** 10-15%

Discount rates should also reflect the risk profile of the evaluating entity. Early-stage biotech companies typically face higher operational and financing risks and should incorporate this uncertainty into their models. In contrast, established pharma companies benefit from greater diversification and financial stability, allowing them to apply lower discount rates.

Table 21 summarizes recommended discount rates based on our consulting experience across various valuation scenarios (Rottgen 2025; Stasior et al. 2018). Early-stage biotech companies conducting internal portfolio assessments or evaluating external acquisitions should apply the stage-specific discount rates previously outlined, with higher rates reserved for earlier-stage and higher-risk assets. Large pharmaceutical companies, however, should use lower discount rates of 10-15% when evaluating acquisition targets or assessing internal development opportunities, reflecting their access to cheaper capital, superior resources, and established drug development expertise.

Similarly, when modeling early biotech companies or assets from an acquirer's perspective, we apply these same lower discount rates of 10-15% to accurately capture the strategic value these opportunities represent to potential acquiring partners. This tiered approach ensures that discount rates appropriately reflect both the inherent technical risk of each asset and the distinct strategic and financial capabilities of the organization conducting the valuation.

DISCOUNT RATES BY COMPANY TYPE AND VALUATION PURPOSE

Company or Use Case	Recommended Discount Rate	Notes
Early-stage biotech companies (for internal portfolio or external acquisitions)	Discovery or Preclinical Stage: 19-28% Phase 1: 18-22% Phase 2: 16-20% Phase 3: 13-17% Registration: 10-15% Approval: 10-15%	Apply the upper bound for early-stage assets with higher capital costs and greater investment risk.
Large pharma companies (for internal or external opportunities)	10-15%	Reflects a lower cost of capital and diversified operations, which is closer to WACC.
Acquirer-side valuation (for early-stage biotech companies or assets)	10-15%	Reflects acquirer perspective, helping biotechs understand how potential partners might value their assets.

Analysis by AGMI Group based on data from Rottgen 2025 and Stasior et al. 2018

Table 21. Suggested discount rates organized by company type and valuation purpose.

The discount rate serves as the fundamental link between future cash flows and present value in Net Present Value (NPV) calculations, represented in Figure 34. By applying the discount rate to expected future revenues and costs, NPV converts these cash flows into today's dollars, accounting for both the time value of money and project-specific risks. In other words, a higher discount rate reduces the present value of future cash flows, resulting in a lower NPV, while a lower discount rate increases NPV. This direct relationship makes the discount rate selection critical for accurate project valuation and informed investment decisions, as even small changes in the discount rate can significantly impact whether a project appears financially attractive.

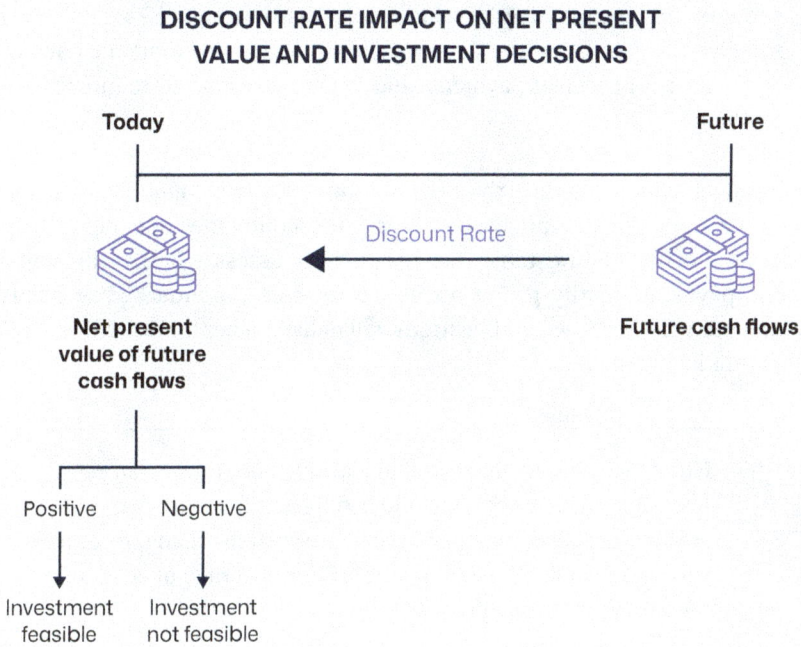

Figure 34. Visual representation of how discount rates impact net present value calculations and investment decisions.

The key lies in calibrating discount rates to realistically reflect uncertainty without excessively discounting future potential. Incorporating stage-specific discount rates tailored to the asset's development phase and the valuation context enables companies to more accurately assess project viability, optimize portfolio strategy, and make better-informed investment and acquisition decisions.

Role of Payment Terms in Licensing Agreements

For early-stage biotech companies, licensing deals is a pathway to monetize intellectual property without bearing the full burden of development and commercialization costs. In these agreements, licensors (early-stage companies or individuals) grant rights to their technology to licensees (usually large firms) (Frei and Dev 2013; O'Connell et al. 2014). The aim is to develop a fair deal structure that accounts for the product's value while distributing risks and rewards appropriately—compensating the licensee for assuming development risks while ensuring the licensor receives payments and remains committed to future success of the product.

Licensing deal structures directly influence how risk and value are distributed between partners, fundamentally shifting valuation outcomes depending on your position at the negotiating table. Licensors must assess whether upfront and milestone payments justify giving up future revenue potential, while licensees must evaluate whether royalty obligations will deliver acceptable returns on their development investment.

> **Note:** For readers interested in learning about university licensing—which operates under different frameworks, stakeholder structures, and incentive systems than corporate-to-corporate deals—see Appendix 5 for coverage of academic technology transfer processes.

The valuation of an asset in the setting of a deal is contingent on several factors: the asset's development stage (preclinical through Phase 3), therapeutic potential, addressable market size, competitive landscape, and the strategic priorities of both parties. By structuring payments across multiple timepoints and contingencies, partners can align incentives while managing their respective risk exposures.

It's important to note that licensing transactions occur in confidential settings with highly specific terms tailored to each situation. Rather than expecting fixed numbers applicable to all scenarios, you should work with industry benchmarks and ranges as starting points for your valuation models, adjusting parameters based on the unique characteristics of your asset and potential partners.

The main components of a deal include upfront payments, milestone payments, royalties, and exclusive agreements. Table 22 provides a summary of their definitions provides practical guidance on how to incorporate each into your valuation models.

LICENSING DEAL STRUCTURE AND VALUATION MODELING FRAMEWORK

Deal Component	What Is It?	How to Model It (Licensee Perspective)
Upfront Payment	A lump sum paid at the beginning of a licensing deal.	Model as an immediate cash inflow in year of deal signing.
Milestone Payments	Payments made when specific development or sales goals are achieved.	Model as one-time inflows at specific achievement points.
Royalties	Ongoing payments based on a percentage of revenue generated.	Model as recurring percentages on annual net sales throughout commercial lifecycle.
Exclusive Agreements	Grants partner sole development and commercialization rights of asset.	No direct cash flow, but enhances value through market control and reduced competition.

Table 22. Overview of key licensing deal components and corresponding financial valuation modeling approaches from the licensee perspective.

Interestingly, analysis of 2,942 licensing deals between 2010 and 2023 reveals that most agreements contain a strategic mix of these components rather than all four, with the specific combination reflecting each deal's unique negotiation dynamics and risk-sharing priorities, seen in Figure 35 (Neuendorf et al. 2023).

**DEAL STRUCTURE COMPONENT FREQUENCY
IN DRUG LICENSING, 2010 TO 2023**

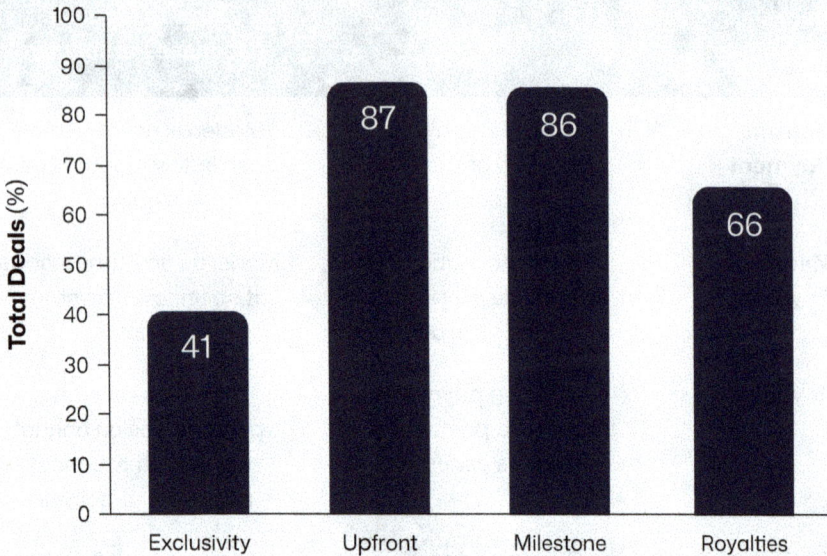

Source: Neuendorf et al. 2023

*Figure 35. Deal structure analysis showing the frequency of key
licensing agreement components of drug licensing
from 2010 to 2023.*

Let's dive deeper into understanding how these payments are impacted by clinical development stage.

Upfront payments are often the most readily available component of a licensing deal due to their simplicity and visibility. These payments typically vary based on the development phase of the asset. A study by Neuendorf et al. described an approach to using upfront payments to predict the total value of licensing deals. Their model predicted the total size of a licensing deal was about 7x that of the upfront size for a Phase 1 deal, 5x for a Phase 2 deal, and 4x for a Phase 3 deal. This suggests that larger upfront payments are associated with assets further along in development, reflecting reduced risk and higher perceived value (Neuendorf et al. 2023).

Milestone payments depend on the reaching of different stages of the product life cycle such as during development (IND, Phase 1, Phase 2, Phase 3, approval, registration) or during the sales period. As the asset progresses through clinical development, the payouts tend to increase, reflecting the confidence of an asset reaching developmental or commercial success. Notably, over 60% of milestone payments are often scheduled to occur post-approval, rewarding successful commercialization efforts and long-term market performance based on sales numbers (Neuendorf et al. 2023).

Royalty rates reflect the value of the intellectual property, with higher rates typically negotiated based on the maturity of the asset, which depends on its progress through clinical development. This is often the most discussed term in the public eye, as it provides long-term payouts without the pressure of meeting specific milestones to receive compensation. Rates vary depending on preclinical and clinical development phases (with increasing rates as the product matures in development, reflecting the reduced risk and increase in value) as well as by type of licensor (i.e., academic institution versus corporation). In addition, royalties are usually "tiered" such that the actual royalty payment owed for a payment period is a function of the annual product sales during that period, multiplied by the negotiated royalty rate for each tier of sales achieved.

Effective Royalty Rates

Given it's becoming more common practice, let's discuss more about how to compare tiered royalty rates. Consider the following example. A license includes an 8% royalty on the first $100M in annual sales, 10% on sales between $100–500M, 15% on sales between $500M and $1B, and 20% on sales above $1B per year. This structure might be broadly characterized as an "8–20% royalty" or a "maximum 20% royalty."

A more accurate way to evaluate the terms is by calculating the Effective Royalty Rate (EFR), the weighted average royalty applied to a specific level of annual sales. In this example, we consider the sales figures of $200M, $500M and $1B. By applying the specific royalty provision to the three assumed sales levels, one obtains a 9.0% EFR for $200M, 9.6% EFR for $500M and 12.3% EFR for $1B in assumed annual sales (see Table 23) (Edwards 2017).

TIERED ROYALTY RATE STRUCTURE: EFFECTIVE ROYALTY RATE CALCULATION EXAMPLE

If $200M in Sales	If $500M in Sales	If $1B in Sales
$8M on $100M	$8M on $100M	$8M on $100M
$10M on $100M	$40M on $400M	$40M on $400M
		$75M on $500M
$18M = 9.0% EFR	$48M = 9.6% EFR	$123M = 12.3% EFR

Source: Edwards 2017

Table 23. Example calculation showing how tiered royalty rate structures result in different EFRs based on total sales volume levels.

A study analyzing royalty rates from 2007 to 2018 found that EFRs are influenced by both the development stage of the product and the type of licensor, such as academic institutions versus corporations, across global and regional markets, see Figure 36 (Edwards 2019). The study found that:

- Worldwide corporate deals exhibit increasing EFR with each successive stage of development, with the largest gains occurring at Phase 3 deal signings.

- Regional corporate deals achieve higher EFRs compared to worldwide deals during the preclinical, Phase 1/2, and Phase 3 stages.

- University deals show minimal EFR gains regardless of deal commencement at more advanced stages of development.

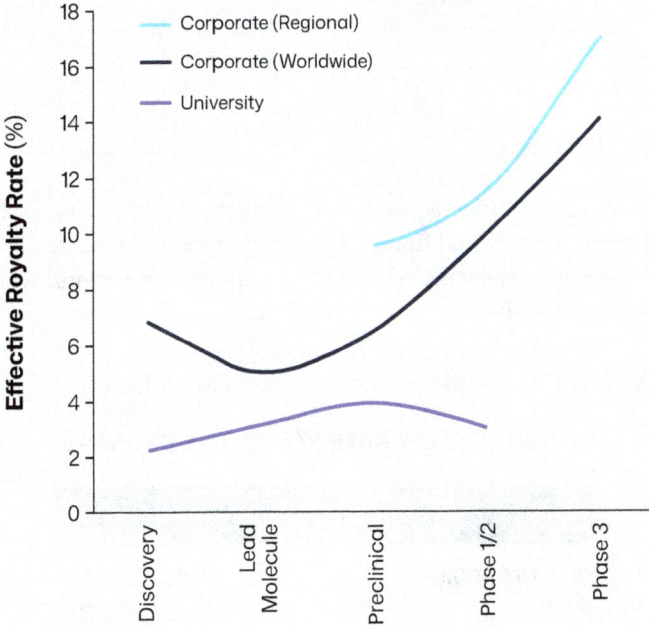

EFFECTIVE ROYALTY RATES BY DEVELOPMENT STAGE AND LICENSOR TYPE, 2007 TO 2018

Legend:
- Corporate (Regional)
- Corporate (Worldwide)
- University

Y-axis: Effective Royalty Rate (%)

X-axis categories: Discovery, Lead Molecule, Preclinical, Phase 1/2, Phase 3

Note: Effective royalty rate (EFR) calculated for assumed annual sales of $500M/year.

Source: Edwards 2019

Figure 36. Pharma licensing market analysis showing average effective royalty rates from 2007 to 2018 across development stages from Discovery and Preclinical Stage through clinical development phases.

In summary, the licensing process demands careful negotiation between the licensor and licensee to balance the various types of payments that can be included in a licensing deal. For a deeper dive into royalties, upfront payments, milestone payments, and their application within the biotech valuation model, please refer to Appendix 2.

Now that you've seen the key components of licensing deals, let's look at how specific deal structures can influence value. In the following examples, we'll explore how changes to royalty tiers, payment timing, and market assumptions can affect financial outcomes for licensors and licensees. These scenarios are meant to show how the details of a deal—not just the totals—can shift who captures more value and when. Keep in mind how each example might feed into the valuation and influence strategic decisions.

Example 1

Tiered Royalties

The details of a negotiated licensing deal can dramatically affect financial outcomes, making it crucial to evaluate different structures carefully. One key factor is how royalties are applied, particularly in tiered royalty agreements, where payout differences can be significant.

Consider a licensing deal with the following structure in Table 24:

TIERED ROYALTY RATE STRUCTURE EXAMPLE

Sales	Royalty Rate
Up to $500M	10%
$500M – $1,000M	12%
Over $1,000M	14%

Table 24. Example of a tiered royalty rate structure showing how rates increase with sales volume thresholds in drug licensing agreements.

At first glance, the differences in percentages may seem small, but as sales grow, the cumulative impact on total royalties can be substantial, especially with tiered royalties. However, this will depend on the terms of the deal.

For example, we can structure royalty rates based on a novel drug's multiple indications. Thus, a deal structure may treat them as either:

1. **Separate Indications:** Royalty rates are applied individually for each indication.

2. **Combined Indications:** All indications are bundled into one agreement under a shared royalty structure.

Table 25 shows the impact of each indication treated separately versus all indications treated as one.

EFFECTIVE ROYALTY RATE COMPARISON: SEPARATE OR TREATED AS ONE

Scenario 1 – Each indication treated separately			
Year 5 Projection	**Sales**	**Rate**	**Payment**
Indication 1	$400M	10%	$40M
Indication 2	$400M	10%	$40M
Indication 3	$400M	10%	$40M
Total	**$1.2B**	**10% (EFR)**	**$120M**

Scenario 2 – All indications treated as one			
Year 5 Projection	**Sales**	**Rate**	**Payment**
	First $500M	10%	$50M
Indications as One	Next $500M	12%	$60M
	Last $200M	14%	$28M
Total	**$1.2B**	**11.5% (EFR)**	**$138M**

Table 25. Comparison of royalty payment outcomes under different deal structures: separate indication versus indications treated as one.

Each approach carries distinct financial implications. By treating the indications separately, each remained below the threshold for higher royalty rates, meaning all sales were subject to the lowest tier of 10%. In contrast, combining the indications into a single agreement pushes total sales past key thresholds, unlocking higher royalty rates and increasing overall payouts.

Since both the licensor (technology owner) and licensee (commercial partner) seek to maximize the value of their deal, it's crucial to analyze the contract details carefully. A well-structured agreement ensures that financial opportunities aren't overlooked, reinforcing the importance of due diligence and strategic negotiation.

Example 2:

Upfront Versus Milestone Payments

One key part of these deals is deciding how much money should be paid upfront versus later, when certain milestones are met. Let's go through a simple example of how adjusting these payments affects the overall value of the deal.

Earlier, we discussed NPV and how it's used to evaluate the value of a project or asset. In biotech, we typically rely on rNPV, which not only estimates the present value of a drug's future earnings but also factors in the probability of success at different stages of development, hence the "risk adjustment."

Table 26 compares the rNPV for both the licensor and licensee under three different deal structures. Each scenario varies in terms of upfront payments and milestone payments, while keeping the royalty rate constant at 9% for a better comparison between scenarios.

DRUG DEVELOPMENT LICENSING: SCENARIO ANALYSIS OF DEAL TERMS

		Scenario		
		#1	**#2**	**#3**
Payment	Upfront	$100M in Year 0	$25M in Year 0	$150M in Year 0
	Milestones	$50M in Year 5	$25M in Years 1-4	None
	Royalty	9%	9%	9%
	rNPV of licensor	$29.2M	$22.5M	$36.1M
	rNPV of licensee	$13.9M	$22.6M	$1.7M
			Better for licensee	**Better for licensor**

Table 26. *Analysis of three licensing deal scenarios showing how different upfront payments, milestone payments, and royalty rates affect rNPV for both licensor and licensee.*

Comparing Licensor and Licensee Shares of rNPV

Using **Scenario 1 as a baseline**, we can see how adjusting upfront and milestone payments affects the financial balance between the licensor and licensee:

- **Scenario 2 benefits the licensee**, as it results in a higher rNPV compared to Scenario 1 and 3. However, this comes with a tradeoff as payments are tied more closely to milestone achievements, which are uncertain and carry execution risks. If milestones are not met, the expected payments will not materialize.

- **Scenario 3 favors the licensor**, as it provides a large upfront payment with no milestone obligations for the licensor. This ensures immediate payout and reduces financial risk for the licensor but loses out on benefits in the longer term (e.g., very high sales revenue).

As illustrated by these three scenarios, each deal is unique, shaped by the strategic goals of both parties. A key consideration is how involved the licensor wants to be in capturing short-term versus long-term benefits. This often requires striking a delicate balance between risk and reward. The longer a licensor remains engaged in a product's clinical development, the greater the risk, but the potential for a larger payout if successful.

In other words, delaying out-licensing (i.e., waiting to license a drug at a more advanced stage) can result in a larger share of the financial upside for the licensor. For example, when licensing a drug that's in the preclinical or IND stage, the licensor might receive 20% of the deal value while the licensee gets 80%. However, for a drug in Phase 2b/3 (as it approaches approval), the split might be more equal, with each party receiving 50% of the deal value (see Table 27). This shows that early-stage licensing shifts more of the clinical development risk onto the licensee, who must be compensated with a larger share of the future financial rewards. Other key factors to consider when determining the split of a licensing deal include development costs, time to market, competitive landscape, and capital requirements.

DEAL VALUE DISTRIBUTION BETWEEN LICENSOR AND LICENSEE BY DEVELOPMENT PHASE

Development Phase	Licensor	Licensee
Preclinical	10~20%	80~90%
IND	20~40%	60~80%
Phase 2b/3	40~60%	40~60%
Approved	60~80%	20~40%

Note: Provided figures are estimates for illustrative purposes only and should not be relied upon for deal negotiations.

Table 27. Comparison of deal value share between licensor and licensee by development phase.

Example 3

Profit-Sharing Between Licensor and Licensee

Lastly, the value of licensing deals is shaped by several key factors. Here, we present different scenarios to illustrate the balance between licensors and licensees, adjusting variables such as the new drug's market share, royalty rates, COGS, and development expenses.

In Table 28, we can see how changing these different scenarios can tip the balance in favor of the licensor or licensee.

LICENSING PROFIT SHARING ANALYSIS: LICENSOR VS. LICENSEE

Case Scenario	Revenue Size ($M)	Licensee share ($M)	Licensor share ($M)	Comment
Base Case	$50	$29	$21	Nearly 50/50 split.
Decrease Market Share by 50%	$13	-$0.83	$13.8	Profits <<< development expenses.
Double Market Share	$125	$88.8	$36.2	Disproportionate benefit from increased sales.
Double Royalty Rate	$50	$14	$36	Licensee shares more of their profits with Licensor.
Reduce Royalty Rate by 50%	$50	$36.5	$13	Licensee retains cash.
Increase COGS from 40% to 50%	$25	$9	$16	Licensee bears 100% of COGS and is disproportionately impacted
Double Development Expenses	$29	$8	$21	Licensee bears 100% of these expenses.

Table 28. Analysis of various scenarios that impact total revenue and how profit is distributed between licensor and licensee in licensing deals.

Doubling the market share or reducing the royalty rate can significantly shrink the licensor's share of the profit. Conversely, cutting the market share in half or doubling the royalty rate allows the licensor to retain a larger share, at the expense of the licensee. Meanwhile, costs related to COGS and asset development tend to disproportionately impact the licensee, who bears the financial risks of drug development.

Remember, deals need to be viewed as partnerships (as with any relationship) and having clear terms that both parties can agree upon is key to the successful development of a biotech product and avoid any conflict or litigations in the future.

What to Keep in Mind

At this point, you've navigated through some of the most practical elements that drive biotech valuation: managing operating expenses (including sales, marketing, and G&A), selecting appropriate discount rates, and structuring licensing deals. These elements may seem technical on the surface—but together, they shape how companies manage their operations, evaluate risk, and negotiate the kinds of deals that determine long-term success.

You now have a toolkit for understanding the commercial infrastructure behind biotech innovation—from estimating sales forces and marketing budgets to applying the right discount rates that reflect both risk and opportunity. You've also seen how the structure of licensing deals can dramatically shift value between partners, and how seemingly small changes in royalty tiers or payment timing can lead to significantly different outcomes.

In the next chapter, we'll bring everything together by applying the SOTP model to a comprehensive biotech valuation case. You'll see how all these components—from R&D costs to revenue projections to SG&A expenses—integrate into a holistic view of a company's product portfolio and overall worth.

BEFORE YOU WRAP UP

Take a moment to reflect on what you've learned about operating expenses, discount rates, and licensing payment structures. These key takeaways and questions are designed to help you connect these concepts to your biotech valuation model and identify areas you'd like to explore further.

Key Takeaways

- Selling, General, and Administrative (SG&A) expenses form the operational infrastructure, evolving from fixed pre-launch expenditure into a proportional relationship with revenue as products gain market traction.

- Planning for asset commercialization begins before product launch, with careful consideration of how sales and marketing resources will scale throughout the product lifecycle and eventually wind down after loss of exclusivity (LOE).

- The discount rate you choose fundamentally shapes how you value future cash flows—higher rates reflect greater uncertainty and risk, while lower rates signal confidence in future returns and can dramatically impact valuation outcomes.

- Seemingly minor differences in deal structures can create major shifts in value distribution between partners, making the specific terms of licensing agreements important to understand and justify.

- The timing of licensing decisions represents a fundamental strategic choice that balances risk against potential reward—early deals push development risk to the licensee and give less future value for the licensor, while waiting to license at later stages gives more potential upside for the licensor at the cost of bearing greater development risk and investment.

BEFORE YOU WRAP UP *(Continued)*

Reflection Questions

1. How might your valuation approach change if you viewed your biotech asset through the eyes of different stakeholders—as the primary developer, licensee, investor, or acquirer?

2. What unique sales and marketing challenges might emerge for therapeutics targeting rare diseases versus those addressing more common conditions, and how would these differences affect your SG&A projections?

3. What factors could influence your selection of discount rates when evaluating biotech assets at different development stages or therapeutic area?

4. What signals about a company's confidence in its technology might be revealed through its chosen deal structures and payment preferences?

5. When might accepting a lower upfront payment in a licensing deal actually result in capturing more value for your company in the long term?

06

Biotech Valuation in Action

After building your foundation—from assessing market opportunities and understanding product dynamics to calculating R&D probabilities and mastering operational considerations throughout the previous chapters—you've now reached the moment where theory transforms into practice. This final chapter is where all the knowledge you've gained comes to life in a tangible, applicable form that will empower your biotech valuation decisions for years to come.

In this chapter, you'll learn how to implement key model variables in Excel to create a practical valuation model you can use immediately in your investment or strategic decision-making process. We'll explore a step-by-step implementation of rNPV models for early-stage biotech assets, a SOTP valuation framework to evaluate multi-asset biotech companies, and a business case that ties everything together, which will serve as a template you can adapt to your own analyses.

By the end of this chapter (and guide), you'll have gained the tools to translate your groundbreaking science into the financial language investors require. This final puzzle piece connects your deep scientific expertise with the valuation methods VCs and pharma partners use when deciding whether to fund or acquire your innovation. The complete picture you'll now see is how your clinical timelines, success probabilities, and market penetration assumptions directly influence your company's valuation in term sheets and partnership discussions. With these practical Excel models in hand, you'll confidently showcase not just the science behind your platform, but also its risk-adjusted value—whether you're defending your pre-money valuation to Series A investors or negotiating milestone structures with potential acquirers. In biotech's high-stakes funding landscape, this financial fluency transforms how stakeholders perceive both you and your company's potential.

What is Discounted Cash Flow (DCF) Analysis?

If you're considering investing time or capital in a biotech product that hasn't yet hit the market, the question is simple: is it worth it? The upside could be huge— but so are the risks. So how do you figure out what it's worth today, long before revenue ever arrives?

This is where DCF analysis comes in handy. Think of it as your financial telescope— it lets you peer into the future and bring those distant earnings back to present day. When investors talk about "intrinsic value," they're referring to what something is actually worth based on the cash it will generate over time, not what the market's mood says it's worth today.

By mapping out a product's future cash flows and then discounting them (applying that discount rate we covered in Chapter 5), you're acknowledging a fundamental truth: money tomorrow isn't worth the same as money today. This approach gives you a rational and data-driven framework that produces consistent valuations across different assets and time periods—exactly what you need when making high-stakes investment decisions.

In the biotech industry, DCF analysis plays a critical role in:

01 **Evaluating biotech and pharma products,** including therapeutic candidates and pipeline opportunities.

02 **Assessing market potential and competitive positioning** for both branded and generic drug manufacturers.

03   **Optimizing asset portfolios** for established biotech and pharmaceutical companies with stable revenue streams.

04   **Facilitating strategic partnerships and contract development,** such as collaborations between biotech startups, large pharmaceutical companies, and CDMOs.

How DCF Analysis Works

DCF analysis typically involves four key steps:

1. **Estimate Future Cash Flows:** Start by forecasting the asset's potential earnings over time. This includes projecting revenue, costs, and timelines— grounded in assumptions about market size, pricing, clinical development, and commercialization. These projections form the foundation of the valuation.

2. **Select an Appropriate Discount Rate:** Choose a discount rate that reflects the non-drug development specific risks involved, including financing, manufacturing, IP, competitive, market uptake, and execution risk. This rate also accounts for the time value of money, helping adjust future cash flows to reflect their value today.

3. **Apply the Discount Rate to Calculate Present Value:** Apply the discount rate to each year of projected cash flow. This step converts future earnings into today's dollars, allowing for a clearer comparison of long-term investments.

4. **Calculate Net Present Value (NPV):** Add the discounted cash flows together. The asset's intrinsic value. This NPV serves as a cornerstone for investment theses, informing go/no-go decisions, valuation negotiations, and capital allocation strategies. Add the discounted cash flows together. The result is the asset's net present value– a single number that reflects what it's worth today, based on what it's expected to generate in the future.

Understanding the Output

NPV represents the present-day value of future cash flows and serves as a core output of DCF analysis.

- A **positive NPV** means the present value of projected earnings exceeds anticipated costs, indicating a potentially profitable opportunity. This suggests the investment is expected to generate more value than it costs, making it an attractive option for investors.

- A **negative NPV** means costs outweigh expected earnings, signaling that the investment may result in a net loss and may be too risky to pursue.

Limitations in Early-Stage Biotech Valuation

DCF analysis faces significant limitations when applied to early-stage biotech companies. Applying DCF to early-stage biotech can be like evaluating a pilot's landing skills before they've learned to take off—the fundamental assumptions required for meaningful analysis may not exist yet. Originally designed for mature companies with stable cash flows, DCF struggles with development-stage biotechs due to three primary challenges:

- **Unstable cash flow projections** from lack of revenue-generating products and unpredictable clinical development costs, timelines, and success rates.

- **Binary event risk** where clinical outcomes and regulatory decisions can dramatically alter valuations overnight.

- **Impact of market sentiment** where hype cycles and broader biotech sector trends significantly influence how investors perceive early-stage biotech prospects and potential success.

Despite these limitations, DCF maintains practical utility when properly contextualized. Rather than seeking precise valuations, use DCF as a structured framework to model ranges of potential outcomes with risk-adjusted probabilities across different development scenarios. The methodology's requirement for explicit assumptions forces rigorous thinking about key value drivers and enables meaningful sensitivity analysis when all inputs, probability assessments, and timelines are properly documented. DCF also provides consistency for comparative analysis across multiple assets or development programs, facilitating relative prioritization and resource allocation decisions even when absolute valuations remain uncertain. Finally, most institutional investors and strategic acquirers continue to expect DCF analysis as part of comprehensive valuation packages—not because it provides perfect answers, but because it offers a common analytical language for investment discussions and due diligence processes. Given these practical benefits alongside the inherent challenges of early-stage development, this is why early-stage biotech valuation uses risk adjustment to address these unique industry challenges.

Approach to Risk-Adjusted Net Present Value (rNPV)

While DCF is a powerful tool, traditional models don't fully account for the unique risks in biotech—especially the uncertainty around clinical trial success. That's where rNPV is useful. rNPV enhances traditional DCF analysis by incorporating the probability of success at each stage of development, preclinical or clinical. In other words, this approach risk-adjusts cash flow projections to present value based on the likelihood of progressing through clinical development and regulatory approval.

Among valuation methods for individual biotech products, rNPV is the most suitable because it accounts for the high uncertainty of preclinical and clinical development while providing a more realistic estimate of an asset's true value. This method is particularly useful for evaluating:

- **preclinical and clinical-stage biotech assets** with uncertain clinical development and approval outcomes
- **innovative and high-risk pharmaceutical and biotech assets** that push the boundaries of science and therefore have minimal to no comparable assets in the market
- **any life sciences assets progressing through phased clinical development**, where success probabilities can be reasonably estimated or are well documented in the industry

Adding rNPV to DCF Analysis

The process follows traditional DCF analysis but adding one key step early in the process between "Estimate Future Cash Flows" and "Select an Appropriate Discount Rate":

2. Risk-Adjust Future Cash Flows: Apply probabilities of success at each clinical development phase to adjust projected cash flows. This step accounts for the likelihood of progressing through regulatory hurdles and provides a more realistic estimate of value by incorporating the attrition risk inherent in drug development.

This ensures that potential earnings reflect the risks associated with biotech R&D. For example, if a product is expected to generate $500M in future revenue but has only a 30% chance of success, the risk-adjusted value would be $150M—reflecting the likelihood that those earnings will actually be realized.

This adjustment is applied across all projected cash flows—revenue, costs, and other financial inputs—at each development stage. Once these risk-adjusted figures are calculated, the remaining DCF steps proceed as usual: discounting future cash flows to present value and summing them to determine the final rNPV.

By incorporating probability of success, rNPV offers a more grounded, data-driven view of what a biotech asset is truly worth today—an essential tool for financial decision-making in a field marked by uncertainty.

For companies with a portfolio of assets at different stages of development, however, a more comprehensive approach is needed to capture total company value.

Approach to Sum-of-the-Parts (SOTP)

The SOTP method values a biotech company by assessing each of its different components—or "parts"—individually, then adding them together to estimate the company's total worth. This approach is especially useful for biotech companies with a diverse portfolio of products at different stages of research, development, and commercialization, each carrying separate financial risks.

The process typically involves three key steps:

Value Each Product Individually

01

Apply rNPV to risk-adjust future cash flows (revenue and expenses) to estimate the value of each drug based on probabilities of success of completing the next clinical development phases until approval.

Calculate Cash Flow of Company-level Financials

02

Apply standard (non-risk-adjusted) DCF analysis to future cash flows of working capital, CapEx, and D&A.

Combine Valuations

03

Sum the individual biotech product valuations as well as company-level financials to determine the company's total worth or market value of equity.

SOTP is particularly useful for evaluating resource allocation, risk exposure, and strategic decision-making within a company's biotech portfolio. For example, if a late-stage therapeutic candidate is assumed to face a regulatory setback, this method ensures that its valuation impact can be appropriately adjusted without disproportionately affecting early-stage assets. In response, the company may:

- reallocate resources to high-potential pipeline candidates or new projects
- adjust its clinical strategy to optimize future development timelines in the company to ensure financial stability
- pursue strategic partnerships to mitigate financial risk and enhance commercialization prospects

Quick tip: When using the SOTP model, it's helpful to compare your company's valuation to a comparable company with a similar R&D profile and treatment category. If your valuation appears significantly higher, it could indicate that some of your assumptions are overly optimistic and may need to be adjusted.

Best Practices and Challenges in Biotech Valuation

· · · · · · · · ·

Biotech valuations can vary significantly depending on the chosen methodology, making it essential to adopt a rigorous and well-validated approach. Therefore, for any asset valuation, it is important to have a sound and validated approach that:

- ⌄ is **unbiased, objective, and data-driven**

- ⌄ utilizes **clearly-stated assumptions**, incorporating both primary and secondary market research

- ⌄ maintains **clarity and practicality**, avoiding unnecessary complexity that could hinder negotiations

Furthermore, it's important to address several key challenges that arise when valuing biotech products and companies. These challenges largely stem from the high-risk nature of product development, long timelines, and the often unpredictable financial outlooks related to probabilities of success in clinical development and regulatory approval.

Keep these factors in mind as you develop your valuation model:

- ⌄ **Lack of Revenue and Negative Cash Flow:** Many biotech companies operate with little or no revenue, instead relying on service agreements, research grants, partnerships, or external investments. As a result, they often experience negative cash flow for extended periods before reaching a product launch.

  Tip: Be realistic about the amount of capital available and avoid manipulating your model to minimize negative cash flow. It is expected there will be negative cash flow during the R&D and pre-commercialization stage.

- ⌄ **Long Time Horizon to Profitability:** Product development and regulatory approval can take more than 10 years, which means it may be a significant amount of time before a company reaches positive cash flow.

  Tip: Clearly outline the time and capital for R&D required to successfully develop your product and lean toward more conservative time estimates to provide a more reliable projection.

⅄ **Uncertain Projected Cash Flows:** The overall probability of an early-stage product successfully reaching the market is low, which makes future revenue projections highly uncertain.

> *Tip: This is where the rNPV in our model becomes essential. It helps you account for the inherent risks and uncertainties of drug development in your valuation via discount rates and risk-adjustment based on probabilities of success in clinical development.*

⅄ **Preclinical Stage Modeling Complexity: Including preclinical stages in valuation models can create difficulties in explaining** rNPV calculations, as assets in preclinical development have not yet established data demonstrating they will treat their intended indication.

> *Tip: For ongoing preclinical assets, model cash flows starting directly at Phase 1 (bypassing the Discovery and Preclinical stages) and apply a very high discount rate to represent the substantial investment risk. Note that these models are highly inaccurate due to the extended development timeline, very low probabilities of success, and reliance on uncertain inputs inherent in very early-stage assets. For additional insights on the challenges of very early-stage valuation, see the section "Why Your Preclinical Asset Shows Negative rNPV (And Why That's Normal)."*

⅄ **Limited Public Comparables or Benchmarks:** Most early-stage biotech companies are privately held, making it challenging to find reliable benchmarks from publicly traded biotech companies that report financials.

> *Tip: When using any benchmarks, always clearly report your sources and assumptions. This will help ensure the integrity of your model. Look for reputable and up-to-date sources such as life science research, business consulting, or academic reports.*

⅄ **Scarcity of Royalty & Milestone Data:** Licensing deals are often not publicly disclosed or are inconsistently reported, making it difficult to incorporate this data into biotech valuations.

> *Tip: Since many deals happen behind closed doors and do not disclose deal details, we typically exclude royalties and milestone payments unless you have reliable information or specifically evaluating their impact on your rNPV.*

See Appendix 3 for guidance on incorporating licensing payment structures into your model.

Why Your Preclinical Asset Shows Negative rNPV (And Why That's Normal)

You've built a sophisticated rNPV model for your promising preclinical asset. The science is solid, the in vivo data looks promising, although not yet complete, and the mechanism makes biological sense. Yet when you run your carefully constructed model, the output delivers a jarring contradiction: negative rNPV.

The harsh reality is that your model isn't evaluating the quality of your science. Instead, it's making a ruthlessly objective assessment of the financial risk-return profile of bringing that science to market. Even exceptional science can represent a poor financial bet when filtered through the lens of development risk, time horizons, and capital requirements.

Your model is grappling with fundamental challenges inherent to very early-stage drug development:

- **Risk multiplication effects** occur when each development stage introduces probabilities of success that compound across multiple phases and extended time periods.

- **Severe time discounting** heavily penalizes revenue streams occurring more than 10 years in the future.

- **Front-loaded cost structures** worsen based on how early in development your asset sits before achieving any possibility of revenue.

- **Historical success rates** show fewer than 5–10% of preclinical assets ultimately reach market.

- **Unreliable modeling assumptions** force educated guesses about dosing, safety, efficacy, patient populations, competitive positioning, and regulatory pathways because critical inputs remain unknown or highly speculative without real-world clinical data.

Given these challenges, the negative rNPV likely reflects the extreme uncertainty and data scarcity typical of assets in the earliest stages of development. The path forward involves providing the data needed to meaningfully shift your model's assumptions.

Successful completion of preclinical studies, demonstration of acceptable safety profiles, and evidence of target engagement can dramatically improve your valuation inputs. Each successful milestone reduces the risk multiplication effect, shortens the time to potential revenue, decreases the front-loaded cost burden, and improves your asset's cumulative success rate.

Rather than forcing premature valuations with inadequate data, focus on achieving the next critical inflection point where meaningful valuation becomes possible. Your asset needs more proof points, not more sophisticated modeling.

CASE STUDY
Building a Valuation Model

Now that we've explored the key components of biotech valuation, let's put it all into practice. Using the rNPV framework, we'll walk through how to build a flexible, risk-adjustable model that can be tailored to a single biotech asset or company with multiple products at various stages of development.

This approach allows you to:

- **Evaluate value**: Assess the financial worth of either a single biotech product or an entire company with a portfolio of assets.
- **Account for risk**: Incorporate risk-adjusted projections that reflect the uncertainty inherent in clinical trials and regulatory approvals.
- **Adapt to changing conditions**: Build a model that responds to shifts in clinical outcomes, policy decisions, and evolving market conditions.
- **Break down pipeline assets**: Value each product individually to understand how it contributes to the overall portfolio and company valuation.

To support this process, we've created a downloadable Excel template that includes all the key variables and calculations you'll need. The model is designed to be intuitive and doesn't require a background in finance—simply enter your product or research data, and it will generate the valuation output automatically. You can download the prebuilt Excel model from the Resources section of this book.

Next, we'll introduce an example biotech case study to apply the concepts you've learned in the previous chapters.

ONC-301 Lung Cancer Biologic

AGMI Therapeutics is a hypothetical early-stage biotech company developing the clinical asset ONC-301, a novel biologic therapy for advanced non-small cell lung cancer (NSCLC), planned to launch on year 2034. ONC-301 targets a unique pathway involved in tumor proliferation and immune evasion, offering potential efficacy advantages over existing checkpoint inhibitors and targeted therapies. The following analysis examines the market potential and key financial projections for this novel asset.

Market Opportunity

- **Disease Background:** NSCLC accounts for ~85% of all lung cancer cases, with significant unmet medical needs in late-stage and refractory patients (American Cancer Society 2024).

- **Incidence and Prevalence:** In the United States, the incidence of lung cancer is 49 new cases per 100,000 people annually, which amounts to approximately 230,000 new diagnoses each year (SEER 2025). The prevalence, representing the total number of people living with lung cancer at a given time, is estimated to be around 600,000 cases. The general population in the United States grows at an annual rate of approximately 0.8 to 1.0% (Wilder 2024).

  - In this example case, we will use prevalence and assume around 510,000 patients have NSCLC (600,000 lung cancer patients * 85% of these patients have NSCLC).

⅄ **Current Treatment Landscape:** Standard treatments include immune checkpoint inhibitors (e.g., Keytruda [pembrolizumab], Opdivo [Nivolumab]), targeted kinase inhibitors, and chemotherapy. Resistance and side effects remain a key issue, creating demand for novel biologics (Keytruda 2025a).

⅄ **Total Addressable Patient Population:**

 ◆ *Segmentation:* Focus on non-metastatic (i.e., localized or regional) disease (~40% of cases; SEER 2025): ~204,000 patients, often treated in combination with radiation, chemotherapy, and surgery.

 ◆ *Treatment, Access, and Adherence/Compliance Rate:* Assume 80% treatment rate (given metastatic NSCLC has treatment rate of 71.4% (Reinmuth et al. 2013) and non-metastatic NSCLC patients are likely healthier and willing to get treatment), 75% access rate, based on insurance coverage versus out-of-pocket (Keytruda 2025b), 85% adherence rate (based on IV infusion rate, see Chapter 2): **104,040 patients**

⅄ **Market Share:** For simplicity purposes, assume ONC-301 is the 3rd entrant in a three-drug market and gains a 30% peak market share based on order of entry among eligible NSCLC patients, the potential patient pool is **104,040** x 30% = **31,212 patients** in the U.S.

Price Assumption: Using Keytruda as a comparable benchmark, the standard dosing regimen is 200 mg every 3 weeks (Ghoshal 2025). For patients with localized disease receiving one year of treatment (Kraus and Schmid 2024), this amounts to 3,467 mg per patient annually. At approximately $55 per mg (list price) (Schouten 2025), with an estimated 20% reduction from discounts and rebates, the net price is approximately $44 per mg. This translates to roughly $150,000 per patient per year for ONC-301, which aligns with the ASP of comparable hematology-oncology drugs that range from $60,000–$280,000 per patient annually in 2025 USD (Kantarjian and Patel 2017).

Back-of-the-envelope Calculation

1. **Current Full Market Opportunity:** (Bullish Assumption)

 Assuming full penetration if the non-metastatic NSCLC market, **104,040 patients** × $150,000 = $15.6B at 100% market uptake in year 2041

2. **Current Partial Market Opportunity:** (Conservative Assumption)

   Assuming 30% market share of TAPP, **31,212 patients** x $150,000 = $4.7B* in unadjusted peak revenue (i.e., not adjusted for risk of asset clinical development)

* This example calculates unadjusted peak revenue without accounting for annual population growth or drug price increases. While simplified, it provides a useful baseline estimate as a starting point for valuation analysis.

Step 0

Orienting to the Excel Model

The example Excel model is designed to help you value individual assets and/or company portfolios. Take some time to look at the tabs included in the model to orient yourself.

The exercise is structured in three main sections to make your analysis straightforward:

- **Assumptions Tab (Your Starting Point):** This tab contains all the key inputs by column for individual assets. It also contains the input for relevant company-level variables if you are doing a SOTP valuation. For example, here you'll input details like date of launch, market share, pricing, and costs for each product.

- **Patient Population Calculator Tab (Your Calculator):** This tab helps you calculate the number of patients with the disease starting from U.S. total population, then using prevalence or incidence data and specific disease characteristics to segment the population.

- **Summary Tab (Your Dashboard):** This tab provides a consolidated view of both individual asset valuations and the overall company valuation. It's your go-to page for seeing the big picture—which drugs drive the most value and what the entire company is worth.

- **Output Tabs (Your Analysis):** These tabs automatically calculate detailed rNPV analyses for individual assets as well as the company overall (if needed). No complex formulas needed on your part; the model does the heavy lifting based on your assumptions.

⅄ **Appendix (Your Reference Guide):** These tabs contain supporting benchmarks for your valuation model, including uptake curves, market share values, COGS percentages, discount rates, and other industry standards. The appendix also includes an optional Monte Carlo simulation (discussed in Appendix 3) to test various market scenarios and sensitivity analyses.

Step 1

Gathering Assumptions

With the relevant background established, we can now gather the key inputs for our valuation model. These include the clinical development timeline (factoring in probabilities of success, costs, and duration), market assessment, sales and marketing projections, other cash flow elements, and the projected drug uptake curve.

The following tables summarize these inputs, serving as the variable assumptions in our valuation of the example asset ONC-301. Structuring data in this way will be invaluable when it is time to build your own model, as it streamlines analysis and ensures consistency. Many of these values are derived from the benchmarks discussed in earlier chapters. Here is a summary of the tables that contain the essential variables for our valuation case:

⅄ **Table 29: Clinical Development Variables** show how long each clinical phase will take, what it will cost, and the chances of success at each stage.

⅄ **Table 30: Market Variables** outlines when we expect to launch, how we'll price the drug, and how many patients could potentially use it.

⅄ **Table 31: Sales and Marketing Variables** details the resources needed to commercialize the drug, including sales team size and marketing expenditures.

⅄ **Table 32: Other Cash Flow Variables** includes essential financial factors like overhead costs, tax rates, and the discount rate we'll use to calculate present value.

⅄ **Table 33: Market Uptake Curve** projects how quickly doctors will adopt the asset over the first 8 years after launch until we reach the anticipated peak market share.

> **Quick tip:** Given that these models are highly sensitive to your inputs, it is essential to rely on accurate and up-to-date benchmarks for the key inputs to make the most out of the model.

ONC-301 CLINICAL DEVELOPMENT VARIABLES AND ASSUMPTIONS

Development Phase	Probability of Success[*]	Probability of Occurrence	Per-Patient Cost[*] (2025 $)	Enrolled Patients Per Trial[*]	Cost per Trial (2025 $M)	Duration (Months)
Phase 1	61.5%	100%	$130K	60	$7.8	24
Phase 2	26.8%	61.5%	$100K	140	$14.0	36
Phase 3	42.7%	16.5%	$120K	300	$36.0	36
Application Approval	85.5%	7.0%	$4.5M	-	-	12
Phase 4 or Market Surveillance	-	-	-	-	$5 per year	Market exclusivity period

Probabilities of success, per-patient cost, and enrollment figures based on historical oncology clinical trial data highlighted in Chapter 4

Table 29. *Summary of key clinical development variables and assumptions for ONC-301 for each phase of drug development.*

ONC-301 MARKET VARIABLES AND ASSUMPTIONS

Variables	Input
Date of Valuation	January 1st, 2025
Launch Year	2034
Years to Peak	8
Year of Patent Expiration or Competition Entry	2045 (assume a market exclusivity of 11 years from launch year to LOE)
Final Market Share After LOE	5%
Annual Revenue Decline after LOE	50%
Total Addressable Patient Population	104,040 patients (includes patient segmentation, treatment rate, access rate, and adherence/compliance rate)
Duration of therapy	1 year for each patient
Patient Population Growth Rate	1% (based on U.S. population)
Peak Market Share	30% (assume 2nd in 3-drug market)
Estimated Drug Price	$150,000 per patient per year (Keytruda used as comparison, after discounts and rebates)
Price Increase Rate	2% (close to inflation rate)
Cost of Goods Sold (% or revenue)	10% (benchmarked as biologic)

Table 30. Summary of key market variables and assumptions for ONC-301.

ONC-301 SALES AND MARKETING VARIABLES AND ASSUMPTIONS

Variables	Input
Number of Sales Representative	100 (specialty use)
Annual Cost per Sales Representative	$300,000 (specialty use)
Compensation Sales Rep Growth	2% (based on inflation rate)
Sales Force Capacity In Pre-Launch Year (% of Fully Staffing)	50%
Marketing Expense (% of Sales Force Cost)	70% (higher marketing cost given will not be first to market)

Table 31. Summary of key sales and marketing variables and assumptions for ONC-301.

ONC-301 OTHER CASH FLOW VARIABLES AND ASSUMPTIONS

Variables	Input
G&A (% of Total Expense)	20% **Note:** Applies during pre-launch period and Years 1-3 only when revenue <$540M
SG&A (% of revenue)	30% **Note:** Applies during Years 1-3 only when unadjusted revenue ≥$540M, and throughout Years 4+
Tax Rate	21% (federal corporate income tax rate in U.S.)
Discount Rate	15% (early-stage biotech company)

Table 32. Summary of other cash flow variables and assumptions related to operational expense, tax rate, and discount rate for ONC-301.

ONC-301 PROJECTED DRUG UPTAKE CURVE

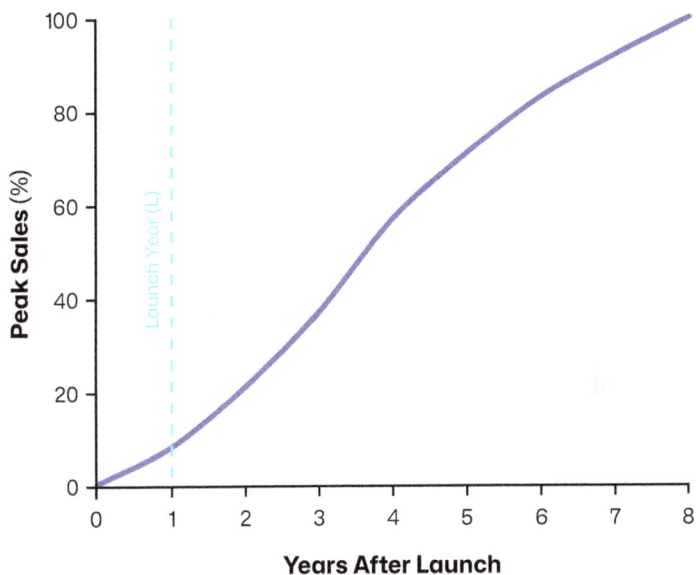

Timeline	Years After Launch	Drug Uptake (%)
Launch Year (L)	1	8
L+1	2	21
L+2	3	37
L+3	4	58
L+4	5	71
L+5	6	83
L+6	7	92
L+7	8	100

Table 33. Projected drug market uptake curve showing adoption trajectory with peak market penetration achieved at 8 years post-launch.

After completing market analysis and gathering the necessary data on the specific biotech industry segment, we can proceed to the next step to prepare our biotech valuation model in Excel.

Preparing the Assumption Table in Excel

A well-structured Assumption Table is the foundation of any biotech valuation model. It acts as a centralized repository for all key variables and assumptions that influence your model output, ensuring consistency and transparency throughout your analysis. By organizing assumptions in one place, you can easily reference, modify, and test different scenarios without disrupting the rest of your model.

In this step, you will set up and populate the Assumption Table using the inputs gathered in Step 1. This structured approach allows for efficient updates and ensures that all calculations remain modifiable as new data becomes available.

Understanding the Assumption Table

In the Excel model, the Assumption Table for five individual assets is located in the "Assumptions" tab. Each column represents an individual asset's parameters and is labeled as Asset 1, Asset 2, Asset 3, etc. Use the "Patient Population Calc" tab to calculate and segment your target patient population by disease characteristics. When an asset targets multiple indications, treat each indication as a separate asset with its own assumptions.

For the ONC-301 valuation exercise, focus on the Asset 1 column. You can ignore the "Other Assumptions" variable section unless you are conducting a company valuation (discussed in Step 4).

Each asset column is generally structured as follows:

1. **General Valuation Assumptions:** This section includes general variables that define the valuation model:

    a. **Date of Valuation:** The date at which the asset is being valued.

    b. **Tax Rate:** The applicable corporate tax rate impacting net income.

    c. **Discount Rate:** The rate used to discount future cash flows to present value.

2. **Asset Assumptions:** This section includes variables about the market such as launch year, market share, pricing, and direct cost assumptions:

    a. **Launch Year:** Expected year the asset will enter the market.

    b. **Years to Peak:** Estimated time to reach peak sales.

    c. **Patent Expiry/Generic Launch Year:** Expected year of patent expiration, loss of market exclusivity, or generic entry.

    d. **Final Market Share After LOE:** Final market share once equilibrium is established after LOE.

    e. **Annual Decline of Revenue After LOE:** Year-to-year percent decline in revenue compared to peak revenue until equilibrium is reached.

    f. **TAPP:** The realistic number of patients who can be effectively reached and treated with a specific therapy after accounting for disease prevalence/incidence, disease segmentation, treatment rate, access rate, and adherence/compliance rate. Applied with annual population growth rate.

    g. **Duration of Therapy:** Expected number of years each patient receives treatment.

    h. **Net Price per Patient Per Year:** The anticipated price of the drug per patient per year. Applied with annual net price increase.

    i. **Peak Market Share:** Expected market penetration based on order of entry, number of drugs in the market, and product differentiation

3. **Manufacturing Expense:** Costs related to the physical production and manufacturing operations of the asset:

    a. **COGS:** Direct costs associated with producing the asset, expressed as a percentage of revenue.

4. **R&D Expense:** This section includes variables that define parameters of clinical development (Phase 1, Phase 2, Phase 3), approval, and post-approval periods:

    a. **Phase 1 to Phase 3:** Includes start date, target enrollment of patients, per-patient cost, cost per trial, duration, probability of occurrence, probability of success, and cumulative probability of success.

    b. **Approval:** Includes start date, probability of success to approval, probability of application approval, cumulative probability of success, application cost, and application duration.

c. **Post-Approval** (Phase 4 or Market Surveillance): Annual R&D expense and duration parameter for Phase 4 or market surveillance activities, including ongoing safety monitoring, real-world evidence collection, and label expansions.

5. **SG&A Expense:** This section includes operational expenses related to commercialization and administrative fees:

a. **Sales:** Annual costs for the deployment and management of the sales force calculated based on number of sales rep, annual cost per sales rep. Applied with annual compensation growth rate.

b. **Marketing:** Annual promotional expenditure expressed as a percentage of sales force costs.

c. **G&A:** Annual corporate and operational costs calculated as a percentage of total expenses during pre-launch years and the first 3 years post-launch.

d. **SG&A After Launch:** Combined sales, marketing, and G&A expenses expressed as a percentage of revenue. Applied during the early commercial period (first 3 years post-launch) when revenue reaches ≥$540M and during the established commercial period (Year 4 onwards).

How to Populate the Assumption Table

1. **Enter Values Gathered during Step 1:** Enter the assumptions you've gathered into the appropriate cells of the Assumption Table under the corresponding asset column. Refer to the "Supporting data" tab for useful benchmark references.

2. **Modify Inputs as Needed:** Refer to the "Model Instructions" tab to identify which cells contain modifiable inputs. Ensure all values are accurate and formatted correctly. Add any citation to benchmarks or data to back up your assumptions.

3. **Check Units and Formatting:** Double-check that each input has the correct units and format (e.g., percentages, currency, or years).

Figure 37 shows an example Assumption Table with three assets. This table mirrors the format provided in the Excel sheet, which you can populate using either the example case data provided or parameters from your own asset valuation.

ASSET VALUATION ASSUMPTION TABLE EXAMPLE

Model Variables	Asset 1	Asset 2	Asset 3
General Valuation Assumptions			
Date of valuation	1/1/2025	1/1/2025	1/1/2025
Tax rate	25%	25%	25%
Discount rate	12%	10%	15%
Asset Assumptions			
Launch year	2030	2028	2032
Years to peak	6	8	5
Year of LOE (patent expiry or competition entry)	2040	2040	2044
Final market share after LOE	8%	5%	10%
Annual decline of revenue after LOE	40%	50%	35%
Patient population with disease (US)	150,000	500,000	75,000
Annual population growth rate	1.0%	1.0%	1.0%
Treatment rate	80%	65%	90%
Access rate	70%	85%	90%
Adherence or compliance rate	75%	85%	70%
Total addressable patient population (TAPP)	63,000	235,000	42,525
Duration of therapy (years)	1	3	2
Price per patient per year (USD)	$85,000	$45,000	$150,000
Price discounts and rebates	0%	10%	20%
Net price per patient per year (USD)	$85,000	$40,500	$120,000
Annual net price increase	2.0%	2.0%	2.0%
Anticipated order of entry	2	1	3
Anticipated # of drugs on market	4	2	5
Peak market share	25.0%	45.0%	18.0%
⋮	⋮	⋮	⋮

Note: Values shown are illustrative examples and do not represent actual asset valuation.

Figure 37. Example assumption table for pharmaceutical asset valuation showing key inputs and assumptions for three drug assets.

> **Quick tip:** If you're unsure about any entry, refer to the provided reference Excel model for guidance. It serves as a complete example to help you verify inputs and formatting.

Examining the Discounted Cash Flow Analysis in Excel

With the Assumption Table completed in Steps 1 and 2 and all relevant inputs incorporated, we can now perform a DCF analysis with rNPV to estimate the value of the biotech asset in question. In the model, you will be able to see future revenues, expenses, and cash flows discounted in real time to their present value.

Understanding the DCF Analysis Structure

The Excel model also includes a complete DCF analysis for each asset, located in separate "Asset X rNPV" tabs (where X represents 1, 2, 3, etc.). Each model uses the corresponding inputs from the "Assumptions" tab and automatically calculates that asset's rNPV. For the ONC-301 exercise, focus on the "Asset 1 rNPV" tab.

Each individual asset's DCF analysis is broken down into the following general sections:

1. **General Information:** The model includes key reference points that define the asset's commercial trajectory:

    a. years to peak

    b. launch year

    c. peak market share

    d. patent expiry/generic launch year

    e. annual sales decline after LOE

2. **Applied Uptake Curve:** Uptake curve describing how quickly the asset is expected to gain market share and grow over time based on time to peak.

3. **Number of Patients on Therapy:** The number of patients on therapy for the year with consideration of general population growth rate.

4. **Revenue Projections** (Risk-adjusted, see methodology box): Revenue forecasts reflect the expected financial returns based on number of patients, uptake curve, TAPP, pricing, and estimated peak market share:

   a. Unadjusted revenue is the projected revenue calculated from patient volume, pricing, and market share assumptions without risk adjustments

   b. Adjusted revenue is the projected revenue that is risk-adjusted based on cumulative probability of success from clinical development to approval

5. **Cost Projections** (Risk-adjusted, see methodology box): Cost projections reflect expenses associated with producing, developing, and commercializing the product:

   a. COGS expressed as a percentage of revenue

   b. R&D based on projected patient enrollment, per-patient cost, trial cost, and duration

   c. SG&A based on anticipated sales force, marketing, and G&A costs

6. **Financial Metrics:** Operating profit is calculated as risk-adjusted revenue minus risk-adjusted costs while taxes are estimated based on applicable local and federal rates and are applied to operating profit.

   a. Net Operating Profit After Tax (NOPAT) is the company's operating profit after deducting taxes

7. **Discount Rate and Factor:** A multiplier used to convert future cash flows to their equivalent present value, using the discount rate.

8. **Present Value of Cash Flow:** Future cash flows are discounted to present value using discount rate and discount factor.

9. **rNPV Output:** The final valuation of the biotech asset is calculated by summing all discounted, risk-adjusted cash flows.

   a. A positive rNPV means the projected cash flows outweigh development costs, suggesting strong financial potential.

b. A negative rNPV suggests the asset's costs exceed its adjusted returns, making it financially unviable unless there are compelling strategic reasons to proceed.

> **Quick tip:** Gross margin represents the percentage of revenue remaining after deducting direct production costs (i.e., COGS). It measures how efficiently a company produces its products and indicates the profitability available to cover operating expenses and generate profit.

Risk Adjustment Methodology

Each revenue and expense category is adjusted according to its specific phase:

- **Clinical Development Phase (Phase 1–3 and approval):** All expenses and any potential revenue (though revenue is typically zero at this stage) must be risk-adjusted by multiplying them by the probability of occurrence and the probability of success for the corresponding phase.

- **Commercialization Phase:** All post-approval expenses and revenue are multiplied by the cumulative probability of success given the entire clinical development pathway has been completed from Phase 1 to approval.

> **Quick tip:** Take your time to review each subsection carefully:
>
> - Ensure accuracy by double-checking inputs from the Assumption Table and corresponding results in the corresponding cash flow.
>
> - Assess whether the results align with expectations based on comparable assets in the same therapeutic category.

Figure 38 illustrates the general structure of DCF analysis, showing how key rNPV components flow through the calculation starting from revenue to valuation output.

DCF ANALYSIS FRAMEWORK: GENERAL STRUCTURE AND COMPONENTS

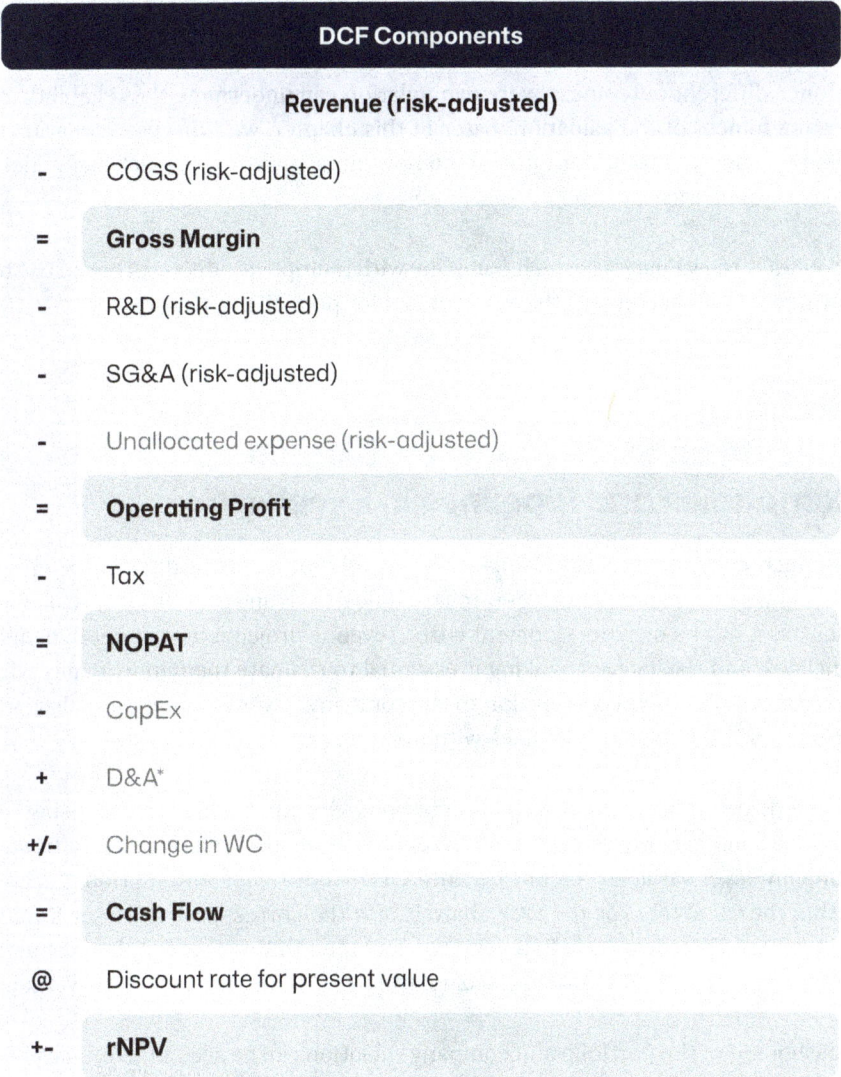

DCF Components
Revenue (risk-adjusted)

-	COGS (risk-adjusted)
=	**Gross Margin**
-	R&D (risk-adjusted)
-	SG&A (risk-adjusted)
-	Unallocated expense (risk-adjusted)
=	**Operating Profit**
-	Tax
=	**NOPAT**
-	CapEx
+	D&A*
+/-	Change in WC
=	**Cash Flow**
@	Discount rate for present value
+-	**rNPV**

D&A is added back because it's a non-cash accounting expense that reduces reported profit but doesn't actually use cash

Note: Grayed out components are specific to company valuations and excluded from individual asset valuations.

Figure 38. General structure of a DCF analysis showing the progression from revenue to rNPV output.

Based on the DCF analysis, the rNPV output for ONC-301 is around +$125M using the inputs from the Assumption Table in Step 2. The positive rNPV suggests that this project is worth investing in given the current inputs, but the results are highly sensitive to changes in assumptions. Since inputs can be adjusted to produce different outcomes, over-manipulation can undermine the reliability and meaningfulness of the valuation. Later in this chapter, we'll discuss scenario and sensitivity analysis to better understand how input variations impact the model in a more systematic way.

If you want to evaluate a biotech company with multiple products, proceed to Step 4, where we will implement the SOTP model for portfolio valuation.

Building an SOTP Model for Biotech Companies and Pipelines in Excel

A biotech company typically manages a portfolio of assets at different stages of development, from preclinical research to clinical trials and eventual market launch. Each asset has its own development costs, revenue projections, market dynamics, timelines, and risk factors, making it essential to evaluate them individually while also considering their contribution to the company's overall valuation. This is why we use a SOTP model for pipelines with multiple assets.

In the Excel model, input key assumptions for up to five assets using the "Assumptions" tab, where each column represents an individual asset. Additionally, company-level variables should be entered in the "Other Assumptions" section within the same tab. For the DCF analysis, use the corresponding "Asset X rNPV" tabs (where X represents 1, 2, 3, etc.) to view the rNPV for each individual asset, and the "Company rNPV" tab to calculate the rNPV of the company-level variables.

A dashboard of the portfolio and company valuation can be seen in "Summary" tab, which lists all key components that contribute to the company's market value of equity. It breaks down the company's value into individual asset's rNPVs, unallocated expenses, and other company-level financial terms (cash flow from CapEx, D&A, and working capital), while also accounting for cash holdings and debt obligations. This format helps you visualize the SOTP valuation approach and understand how each asset and company components contribute to the company's overall worth. Two important calculations to note are the enterprise value and the equity value:

- **Enterprise Value (EV)** represents the value of the operating business to all capital providers (equity holders and debt holders)
- **Value of Equity** is the value attributable to shareholders and is calculated by adjusting the enterprise value for the company's cash and debt—specifically, equity value = enterprise value + cash – debt.

Use this section as a hands-on opportunity to model multiple biotech assets and see how each one contributes to the overall valuation at a company-level. This exercise reinforces how asset-level analysis feeds into broader company strategy and investor decision-making.

Here's how you can proceed with the SOTP model in this exercise:

1. **Navigate to the Excel file** where you'll see the relevant tabs for the SOTP model, including the "Assumptions" tab for inputs of Assets 1, 2, 3, etc., the "Asset X rNPV" tabs (where X represents 1, 2, 3, etc.) for DCF analysis of each asset, the "Company rNPV" tab for DCF analysis of company-level variables, and the "Summary" tab that consolidates all outputs.

2. **Complete Individual Asset Analyses** by analyzing each asset in the pipeline separately. For each asset, you'll need to:

   - collect asset-specific variables and complete an Assumption Table (Steps 1 and 2)
   - generate a corresponding DCF analysis for each asset (Step 3)
   - review the resulting individual rNPV calculations

3. **Include Company-Specific Terms** as discussed in Chapter 5 into the total cash flow of each biotech asset. The variables to consider are:

   - initial CapEx value and CapEx growth rate (typically aligned with the inflation rate)
   - D&A as a percentage of CapEx (generally around 95%)
   - initial working capital value and working capital growth rate (also close to inflation rate)

4. **Include Unallocated Expenses** that account for costs that aren't directly tied to any one product but are necessary for the company's operations and overhead. While these expenses aren't attributed to any specific asset, they're required for keeping the business running and need to be accounted for in the overall company valuation model.

5. **Include Current Company Cash and Debt:** The amount of company cash currently available and the total amount of debt currently owed.

6. **Observe the Automatic Aggregation** as the model adds the individual asset rNPVs to calculate the total pipeline value.

7. **Assess the Total Pipeline Portfolio and Company Valuation** by examining how each asset's contribution to the rNPV affects the company's risk profiles, timelines, and potential returns. This visualization helps you assess the company's projected valuation in total, which will be important for evaluating the company's effectiveness in managing its resources and investments across its product pipeline.

Figure 39 demonstrates an illustrative company valuation, showing how three individual pipeline assets combine with company-level variables to generate the company's enterprise value and value of equity.

EXAMPLE COMPANY SUM-OF-THE-PARTS
VALUATION WITH THREE-ASSET PIPELINE

Component	rNPV
Asset 1	$245,000,000
Asset 2	($12,500,000)
Asset 3	$95,000,000
Unallocated Expenses	($25,000,000)
Capex	($5,000,000)
D&A	$2,000,000
Change in Net WC	($7,000,000)
Enterprise Value	$292,500,000
Cash	$70,000,000
Total Debt	($15,000,000)
Value of Equity	$347,500,000

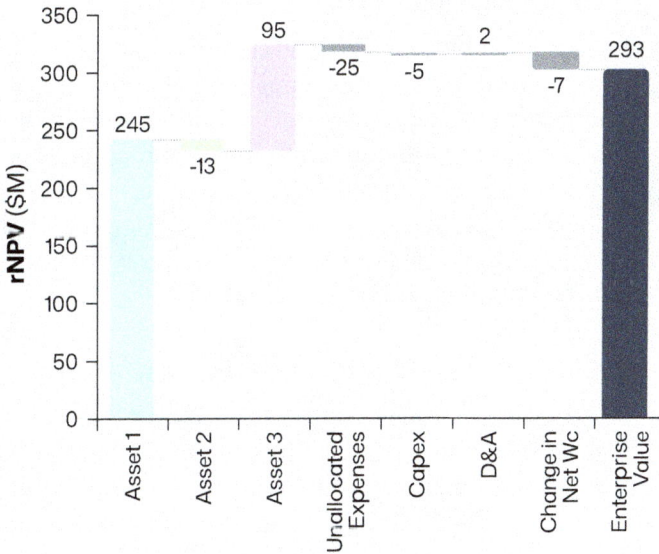

Note: Values shown are illustrative examples and do not represent actual asset valuation.

Figure 39. Illustrative company valuation using SOTP methodology, demonstrating the progression from three individual asset values through corporate-level adjustments to final equity value.

By adding each biotech asset valuation in this step, you'll be able to see where the company's value is coming from, whether it's from a few high potential "blockbuster" therapeutics or a more diversified pipeline with multiple smaller, but promising, products. This approach also ensures a more accurate assessment because each asset has its own clinical development trajectory, market landscape, and probability of success. Ultimately, this helps guide strategic decisions on resource allocation, identifying which products need additional funding or focus, and evaluating the company's overall financial health.

Key Considerations for Valuation Models

A well-thought-out valuation model relies on accuracy and transparency to provide a realistic estimate of a product's or company's value, empowering investors, analysts, and stakeholders to make informed decisions. Keep these important elements in mind:

List Key Assumptions

01

Clearly identify and outline all critical assumptions that drive your model, such as patient population, pricing strategy, market size/adoption, regulatory timelines, and anticipated erosion/generic penetration. Remember assumptions should be grounded in industry norms, scientific data, and competitive benchmarks to enhance credibility. Additionally, reviewing case studies of analogous biotech products can also provide valuable real-world context and strengthen your rationale.

Be as Detailed as Possible for Major Cost Items

02

For significant expense categories such as R&D and sales force, greater granularity will reduce valuation uncertainty. A common pitfall is underestimating R&D costs, so ensure realistic cost modeling based on comparable development programs.

Consider Timing Carefully

03

Avoid assuming accelerated development timelines as your base case scenario or scheduling clinical trials too close together. Realistic timelines based on regulatory requirements and operational constraints are essential for accurate cash flow projections.

Consider Market Size and Any Advantages

04

If your drug targets a rare disease, the business model changes significantly. Orphan drugs often receive pricing advantages, tax benefits, and extended exclusivity, all of which impact valuation.

Document Sources

05

Provide detailed documentation of the sources for each assumption, whether they come from clinical trial data, investor reports, analyst research, or expert opinions. This transparency allows others to scrutinize and validate your inputs, strengthening confidence in your valuation.

Evaluate Valuation Purpose

06

Your model should be tailored to its intended use, whether for investment, partnerships, equity holders, licensing, or internal business planning.

Use Correct Cash Flow Calculations

07

Don't rely on profit metrics (like EBITDA or net profit) as a substitute for cash flow. Adjust for taxes and working capital to compute true free cash flow.

Update Regularly

08

Valuation is highly dynamic, influenced by new clinical trial results, regulatory decisions, competitor developments, and evolving market conditions. Regularly revisit and update assumptions to reflect the most current data, ensuring that your model remains relevant and accurate.

Unlocking Insights with Scenario and Sensitivity Analysis

. . . .

Scenario and sensitivity analysis are essential for evaluating how changes in key assumptions impact biotech valuation, particularly in an industry shaped by regulatory approvals, reimbursement decisions, and competitive dynamics. By systematically adjusting critical variables, such as probability of success, peak market share, pricing, and commercialization timelines, you can assess potential risks and opportunities. This approach not only enhances decision-making but also provides a clearer picture of how uncertainties, like clinical success rates or shifting market conditions, influence valuation outcomes.

Here are the key steps to performing a general scenario and sensitivity analysis:

1. **Base Case:** Start with your base case scenario using the initial assumptions. This serves as the benchmark against which all other scenarios will be compared. Ensure that the base case reflects the average expectations for the biotech product or company, including market size, probability of success, revenue projections, costs, and discount rates. This scenario provides a reference point to evaluate the impacts of changes in assumptions.

2. **Scenario Analysis:**

   * **Best-Case Scenario:** Develop a scenario where all favorable conditions are met. For instance, assume higher market penetration, lower clinical development costs, faster regulatory approvals, and higher probabilities of success.

   * **Worst-Case Scenario:** Conversely, create a scenario with the most unfavorable conditions. Assume lower market penetration, higher development costs, delays in regulatory approvals, and lower probabilities of success.

   * **Moderate Scenarios:** In addition to extreme best and worst-case scenarios, consider intermediate scenarios that reflect more probable outcomes. This provides a broader view of potential results and helps in planning for many possibilities.

3. **Sensitivity Analysis:**

   ● Identify key assumptions that have the most significant impact on the valuation, such as probabilities of success, time duration, market penetration rates, and cost estimates.

   ● Vary these key assumptions systematically and one-by-one to see how these changes affect the valuation. For example, adjust the probability of success for one of the phases of clinical trials by ±10% and observe the impact on the rNPV.

   ● Plot the results such as a histogram to visualize the sensitivity of the valuation to changes in each assumption.

Now, let's consider our biotech valuation model. As you play around with the model, keep these key scenarios in mind. In each case, we isolate the impact of changing one variable while holding all others at their base-case values:

1. **Progression Through Development Stages**: Under the rNPV method, a drug's value increases each successful step it completes during clinical development. Once a phase is completed, the associated R&D costs, time, and probabilities of success are no longer value-relevant—meaning the model updates to reflect only future uncertainty. For example, if an asset already completed Phase 1, the probability of success for Phase 1 is no longer relevant.

2. **Impact of Sales Projections**: Sales forecasts play a pivotal role in rNPV calculations. A drug's revenue trajectory typically follows a pattern: slow initial sales, rapid growth over 6-7 years, stabilization, and eventual decline as LOE and patent expiration come into play. If early sales growth is expected to be strong, rNPV can be significantly higher than under a more conservative projection, even if first-year sales are modest.

3. **Effect of Development Timelines**: Extending the length of clinical trial phases or regulatory review delays commercialization, reducing the drug candidate's value at every stage under rNPV. Since rNPV is based on discounted future cash flows, a longer development timeline increases the discounting effect, lowering the present value of the asset.

4. **Impact of Changing Phase Success Probabilities**: Adjusting success probabilities can significantly impact rNPV at every stage of development.

  - **Increasing Success Probabilities:** Higher probabilities of success boost rNPV, especially in earlier phases where risk is highest. For example, oncology products tend to have higher approval rates than others. Adjust your valuations accordingly to reflect the unique risks and opportunities of the drug's specific disease areas.

  - **Decreasing Success Probabilities:** Lower probabilities of success reduce rNPV, highlighting increased risk and potential hurdles in clinical development. Factors such as regulatory challenges, unexpected safety concerns, or emerging competitive threats may necessitate downward adjustments in valuation to account for these risks.

5. **Increase the Discount Rate or WACC**: A higher discount rate reduces rNPV by increasing the discounting effect on future cash flows, which accounts for higher risk. This effect is most pronounced in early development, when risk is high and projected revenues are farther in the future.

Additional Considerations

For a more comprehensive analysis, consider incorporating a Monte Carlo simulation into your biotech valuation model for sensitivity analysis. This approach uses probabilistic modeling to account for uncertainty by running hundreds or thousands of simulations based on specific variables, providing a more nuanced understanding of potential valuation outcomes. More details on implementing Monte Carlo simulations can be found in Appendix 3.

Additionally, various alternative biotech valuation models may offer different perspectives depending on your specific asset or business case. These models can provide insights tailored to different stages of development, risk profiles, or market conditions. For a deeper exploration of these methodologies, please refer to Appendix 4.

Turning Insights into Action

You've conducted a scenario and sensitivity analysis, adjusting key variables to see how they impact your model's output. Now, what do you do with these insights?

Here are some useful ways to apply them:

Risk Management

Identify the most sensitive variables and prioritize risk mitigation strategies. For example, if your model is highly sensitive to the probability of success in clinical trials, you might focus on improving trial design, enhancing patient recruitment strategies, or diversifying your pipeline to spread risk.

Investment Decisions

Use sensitivity analysis to understand the range of potential outcomes and the factors that drive returns. This insight will help convince stakeholders you have thoroughly thought through all scenarios and set appropriate risk premiums, which will help assess whether an opportunity aligns with their risk tolerance.

Strategic Planning

Scenario and sensitivity analysis allows companies to prepare for different market conditions and develop contingency plans. If a worst-case scenario suggests financial strain, leadership can proactively explore cost-cutting measures, alternative funding strategies, or strategic partnerships to maintain stability.

Stakeholder Communication

Clear, transparent communication fosters trust with investors, regulatory bodies, and internal teams. By presenting a well-supported, data-driven approach to risk management, you demonstrate credibility and strategic foresight. Aligning your team around shared goals and informed decision-making strengthens collaboration and sets the foundation for long-term success.

What to Keep in Mind

In this final chapter, you've discovered how to build valuation models that work for both single assets and entire biotech companies with multiple products in development. You've seen firsthand how a well-structured assumption table becomes your modeling foundation—a place where all your key variables live and can be tweaked as new information emerges. Walking through our example case together, you've gotten hands-on experience evaluating market potential, weighing probabilities of success in clinical development, mapping out revenue streams with DCF analysis, and accounting for various expenditures from development to administrative costs—all within the Excel model you can continue using for your own projects. And perhaps most importantly, you've explored how testing your assumptions through sensitivity analysis helps ensure your model reflects the real-world uncertainty that makes biotech both challenging and exciting.

You've now got the tools to build a biotech valuation model that does more than just run the numbers—it helps you make smarter decisions. Whether you're evaluating a single product or an entire portfolio, you can now structure a model that reflects reality and adjusts as things change.In practical terms, this means:

- you can justify your assumptions with confidence when presenting to stakeholders, investors, or partners

- you know how to adjust your model as clinical data emerges, timelines shift, or market conditions evolve

- you're no longer reliant on industry averages or competitors' approaches—you can run your own numbers and engage in evidence-based strategic discussions

- you understand that valuation is dynamic—something you revisit and refine as science advances, markets shift, and your company strategy develops

This approach transforms valuation from a one-time calculation into a powerful, ongoing decision-making tool that grows with your understanding of both your assets and the broader biotech landscape.

BEFORE YOU WRAP UP

Take a moment to reflect on what you've learned about biotech valuation modeling. These key takeaways and questions are designed to help you connect these concepts to your practical application and identify areas you'd like to explore further.

Key Takeaways

- The risk-adjusted Net Present Value (rNPV) approach is particularly well-suited for biotech assets because it incorporates the unique risks associated with clinical development through risk-adjustment of cash flows based on probability of success.

- A well-structured, centralized repository of key variables ensures consistency across your analysis and allows for quick sensitivity testing as new information emerges.

- As a product advances through development phases, its valuation typically increases as risk diminishes—earlier-stage assets have their cash flows more heavily discounted to reflect higher uncertainty.

- The Sum-of-the-Parts (SOTP) approach demonstrates how companies with multiple assets in a pipeline at different development stages can balance high-risk, high-reward early-stage candidates with more stable late-stage or commercialized products.

- Market conditions, competitive landscapes, regulatory environments, and clinical data all evolve over time, requiring regular model updates with data-driven assumptions to maintain up-to-date and accurate.

- Small changes in inputs can dramatically impact the valuation output, especially for early-stage assets, highlighting the important of utilizing industry benchmarks and expert insights.

BEFORE YOU WRAP UP *(Continued)*

Reflection Questions

1. How might your assumptions about market size, pricing, and uptake curve differ based on your specific therapeutic area and patient population? What sources could validate these assumptions?

2. What specific factors might increase or decrease the probability of success for your asset compared to industry averages for similar therapeutics? Consider mechanism novelty, preclinical data strength, and clinical trial design.

3. How would your valuation change if development timelines extended by 1–2 years for each phase?

4. Which variables in your model have the greatest impact on final valuation when adjusted?

5. If creating a portfolio valuation using SOTP, how would you balance investments across early-stage (high-risk/high-reward) versus late-stage (lower-risk/lower-reward) assets to optimize overall company value?

6. How might you adjust your discount rate to reflect specific non-drug development beyond what's captured in probabilities of success in clinical development?

A Final Word

We've reached the conclusion of our biotech valuation journey. If we were sitting across from each other right now, coffee cups in hand, I'd want to make sure you walk away with not just formulas and frameworks but with confidence in your ability to apply them.

Throughout this guide, we've dissected everything from identifying untapped market opportunities to modeling drug launches, while navigating the probabilities of R&D success that make biotech valuation such a fascinating and unique challenge. But let's be clear about something: the spreadsheets are just tools. What really matters is your judgment—your ability to question assumptions, recognize patterns, and adjust when the data doesn't quite align with the real world.

The biotech industry is defined by its volatility, where unexpected scientific breakthroughs such as gene therapies and RNA technologies can suddenly emerge, promising clinical trials either succeed spectacularly or falter disappointingly, and regulatory environments transform dramatically through measures that cap prescription drug price—all creating both opportunities and challenges for investors and companies alike. This is precisely why careful valuation matters so much. It is during these periods of industry turbulence—particularly those affecting public health during pandemics, economic crises, large shifts in federal policies— that your analytical framework becomes an invaluable compass for navigating uncertainty and identifying opportunity.

What comes next for you? Perhaps its building your first model for a promising therapeutic, or maybe it's revisiting valuations you've already completed with fresh insights. Either way, don't work in isolation. Connect with medical experts, follow updates from the FDA, participate in forums where industry professionals share methodologies for improved their own valuation models.

If you are feeling really brave about valuation and need more practice, the book Valuation Workbook: Step-by-Step Exercises and Tests to Help You Master Valuation by McKinsey & Company has great exercises to test and expand your knowledge.

We've shared everything we believe you need to approach biotech valuation with clarity and honesty, but the learning doesn't stop here. The most valuable models evolve as you gain experience and incorporate new information. Trust your process but never stop refining it.

All the best in your future work. We have a feeling you're going to do remarkable things with what we've explored together!

Acknowledgements

B ringing this book to life required the exceptional support and expertise of many colleagues and collaborators. We extend our deepest gratitude to our AGMI Group team, beginning with Annie Chen, Management Consultant, for her instrumental role in developing the comprehensive biotech valuation Excel model that serves as the practical foundation of this guide. Guan-Lun Liao, Co-founder and U.S. Managing Partner, and Annie Lin, Chief of Staff, provided invaluable assistance, ensuring model integrity through rigorous testing of both functionality and design. Management Consultants Nathan Darius and Chris Chen contributed essential real-world validation by applying our framework to real-world biotech valuation cases and offering critical feedback that enhanced the model's practical utility. Our appreciation also goes to Raghav Ramabadran, Global Partner of Private Capital and Life Sciences, whose forecasting model offered insights that helped us incorporate the best practices in asset valuation and due diligence methodologies.

Special thanks go to Sam Lin, Vice President of Commercial Operations at PharmaEssentia, who generously shared his deep knowledge of gross-to-net discount dynamics, enriching our understanding of pharmaceutical pricing complexities. We are also grateful to our exceptional publishing team: Jess Lomas from It Was Good for her meticulous editing and structural guidance, Ning Sen from NINGs.studio for the compelling cover design, and Farhan Shahid and his talented team from Andromeda GFX for their seamless interior layout and rapid turnaround.

Finally, we acknowledge the countless biotech professionals, investors, and researchers whose work forms the empirical foundation upon which this guide is built—your contributions to the field make resources like this possible.

Appendices

T hese appendices provide additional content that complements the core concepts covered in the main chapters. They offer deeper insights and practical frameworks to support the application of biotech valuation methods in real-world contexts. While not essential to grasping the fundamentals, these materials serve as valuable references for readers looking to refine their analytical approach or explore specialized topics.

APPENDIX 1

Gross-to-Net Deductions and Their Impact on Product Lifecycle

> This appendix expands on key concepts from Chapters 2 and 3, offering a deeper look into gross-to-net deductions and their impact on actual revenue received by manufacturers. Use this section as an additional reference when estimating real-world margins in the complex U.S. drug distribution system, drug pricing, and changes to the product lifecycle in your valuation model.

In the modern U.S. pricing environment, biotech and pharma manufacturers must carefully consider gross-to-net dynamics early on in their commercial strategies, particularly when forecasting revenue for a new drug launch. As discussed in Chapter 2, gross-to-net represents the difference between the gross price (referred to as the WAC) and the list price received by the manufacturer. This gap arises due to a complex web of mandatory and negotiated discounts, rebates, fees, chargebacks, etc., all of which are influenced by regulations, market trends, and payer dynamics.

To clarify, the gross revenue is calculated by multiplying the WAC price by the number of units sold, while net revenue is what remains after all deductions are applied. This gross-to-net difference is estimated to be around 45% (Rottgen 2025), although varies significantly based on non-disclosed negotiations between stakeholders, such as pharma companies, PBMs, insurers, and CMS. The ASP in relation to the list price is a related but distinct term more relevant in Medicare reimbursement calculations, representing the volume-weighted average price of a drug across all purchasers, net of most discounts and rebates.

Over the years, the gross-to-net disparity has widened. While list prices for brand-name drugs have steadily increased, net prices have declined due to growing gross-to-net deductions. In fact, the total estimated value of these deductions across the industry reached between $220B and $260B in 2022, which is an approximate 33% increase from 2018 (Fein 2023a). For manufacturers, understanding these evolving gross-to-net trends is crucial, as they can significantly impact a product's market access and financial performance.

Given the complexity of gross-to-net calculations, this section does not aim to provide a precise formula or modeling strategy. Instead, we offer an introduction to key factors that influence gross-to-net as well as its impact on the product life cycle, which is relevant for early-stage biotech valuation modeling.

Drug Distribution System in U.S. Healthcare

Let's start with an overview of the U.S. drug distribution system. As illustrated in Figure 40, the flow of prescription drugs, services, and payments among various stakeholders in the distribution system is highly intricate, particularly for drugs covered under private insurance (Sood et al. 2017). This complexity, involving manufacturers, wholesalers, pharmacies, PBMs, health plans, plan sponsors, and beneficiaries, makes it challenging to accurately predict gross-to-net outcomes.

U.S. HEALTHCARE SYSTEM: STAKEHOLDER
RELATIONSHIPS AND INTERACTIONS

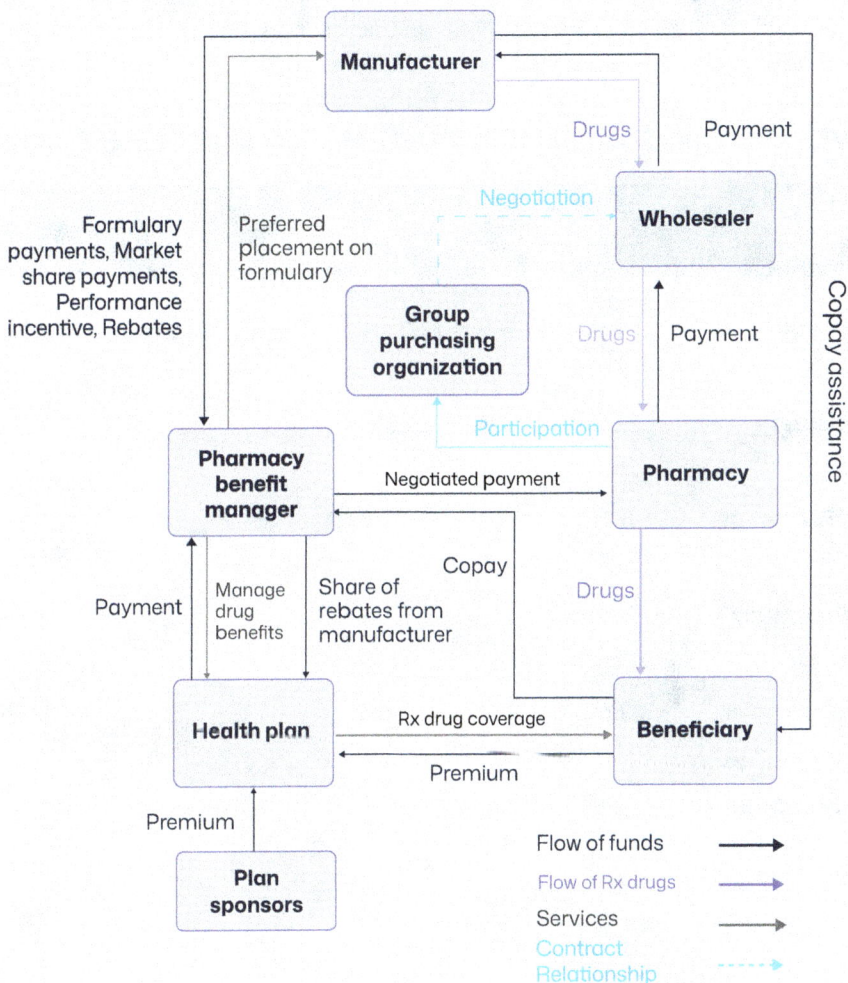

Figure 40. *Relationships and interactions between key parties in the U.S. healthcare system.*

Source: Sood et al. 2017

Additionally, each entity within the pharmaceutical distribution chain operates with its own incentives as well as gross and net margin goals. A study analyzed 2015 company filings and calculated the distribution of average gross margins to reveal how money flows through the distribution system for branded and generic drugs. Table 34 highlights average gross and net margins across all drugs, as well as branded and generic-only categories (Sood et al. 2017).

AVERAGE GROSS AND NET PROFIT MARGINS IN THE U.S., 2015

	All drugs		Brand only		Generic only	
Sector	Gross	Net	Gross	Net	Gross	Net
Insurer	22.2%	3.0%	22.2%	-	22.2%	-
PBM	6.3%	2.3%	2.0%	-	8.0%	-
Pharmacy	20.1%	4.0%	3.5%	-	42.7%	-
Wholesaler	3.7%	0.5%	1.0%	-	18.5%	-
Manufacturer	71.1%	26.3%	76.3%	28.1%	49.8%	18.2%

Source: Sood et al. 2017

Table 34. Average gross and net profit margins for individual companies across U.S. industries in 2015.

In the branded drug market, manufacturers capture the highest gross margins at 76%, while other distribution players see overall lower gross margins, up to 22% (Sood et al. 2017). In contrast, the generic drug market shifts profitability toward distributors, with pharmacies at the highest, enjoying particularly high gross margins of 43%, and manufacturers dropping significantly to 50%. Net margins by brand or generic are only available for manufacturers, which fall significantly below their gross margins at 28% and 18% respectively. This results in manufacturers capitalizing on brand-name profits but distributors benefiting more from generics, especially pharmacies.

Another way to look at this data is that for branded drugs, manufacturers earn about three times more gross profit compared to generics. However, the dynamics for generics are quite the opposite with PBMs seeing a four-fold increase in gross profits from generics compared to brands, while wholesalers make 11 times more, and pharmacies earn almost 12 times as much. Crazy!

Key Factors Influencing Gross-to-Net

There are several factors that influence the gross-to-net calculation. These factors can vary depending on the specific product, market, and regulatory environment. Some of these key factors that play a role include:

1. **Rebates to payers** are often the largest component of the non-mandatory deductions in a drug's gross-to-net calculation, which includes negotiated (for all payers) and statutory (for government payers) rebates.

   a. *Negotiated rebates* are offered to secure formulary placement, manage access restrictions, and reduce patient cost-sharing. These rebates are one of the most critical components for biotech or pharmaceutical manufacturers seeking to improve broader patient access.

   b. *Statutory rebates* are required for government payer programs, such as Medicaid and Medicare. Medicaid plans typically offer the best rebates, but Medicare is becoming increasingly important. Recently, the Inflation Reduction Act introduced rebates for drug list price increases, which outpaced inflation as well as increased manufacturers' liabilities for high-cost Part D drugs.

2. **Fees and discounts to channel participants** include payments to wholesalers, group purchasing organizations, and other distribution partners.

   a. These costs may vary depending on the manufacturers' distribution strategy and can exceed 10% of the gross price or WAC (Seeley 2022).

3. **340B discounts** require manufacturers to provide significant drug discounts to eligible hospitals and clinics serving low-income and uninsured patients. This is the largest provider-focused contributor to a drug's gross-to-net calculation.

   a. The program has recently expanded significantly, reaching $52B in 2022, which accounts for approximately 20% of total manufacturer gross-to-net (Fein 2023b). Product exposure to the 340B program can vary that will require detailed market analysis, as non-340B discounts can be more substantial for consolidated or integrated providers with greater leverage in comparison.

4. **Patient assistance programs** help reduce out-of-pocket costs for commercially insured patients, improving medication affordability and adherence.

   a. These programs are designed to offset commercial patient cost exposure but can be a substantial expense, requiring careful design and implemented by manufacturers.

5. **Asset Therapeutic area** can play a role in the magnitude of an asset's total gross-to-net, as seen in Figure 41 (Guth et al. 2024).

   a. For innovative assets approved by the FDA in 2016 to 2020, gross-to-net was lower in the therapeutic area of oncology, reflecting the clinical severity, high unmet need, and "protected class" status of these treatments. These factors reduce payers' leverage and motivation to negotiate discounts. Similarly, orphan diseases tend to have low gross-to-net due to limited competition, significant unmet need, and relatively low overall spending. As a result, non-oncology therapeutic areas with orphan designation also exhibit lower gross-to-net.

   b. More specific therapeutic areas such CNS or infectious disease did not show clear gross-to-net trends. This likely means that individual assets and markets are likely to be larger drivers in gross-to-net including site of care, insurance benefit, payer mix, and competition.

GROSS-TO-NET DEDUCTIONS BY THERAPEUTIC AREA, 2022

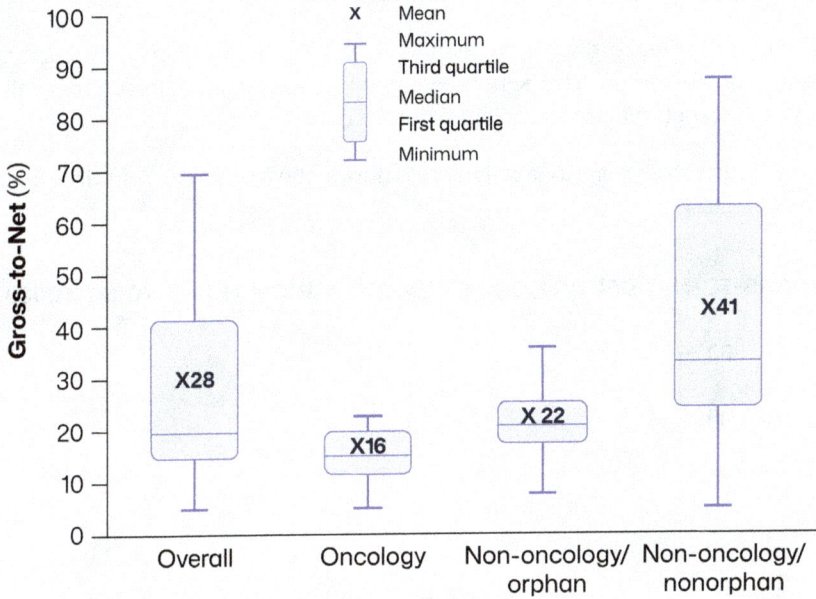

Note: Data includes drug launches from 2016 to 2020 with gross-to-net data from 2022.

Source: Guth et al. 2024

Figure 41. Gross-to-net deductions in 2022 by therapeutic area for 94 innovative asset that launched from 2016 to 2020.

Gross-to-Net Impact on the Product Life Cycle

Launching a new biotech asset or drug is critical to long-term financial performance and depends heavily on the manufacturer's market access and pricing strategy, as discussed in Chapter 2 on market opportunity. The first few years post-launch are especially important for gaining traction among early adopters, which can significantly influence total revenue, as explored in Chapter 3 on launch curves.

Given the importance of this early period, manufacturers tend to be less aggressive in payer rebates and formulary negotiations to ensure sufficient demand and access for providers and patients. For example, investing in patient assistance programs can help maximize early uptake by generating initial demand while broad payer coverage is still being established. These programs may include covering the full cost of the product for patients who are not yet insured, which understandably reduces the net price per patient in the short term.

However, the primary goal is to reduce access barriers, build early customer demand, and foster positive sentiment among patients and prescribers—all of which can steepen the product's initial uptake curve.

Figure 42 illustrates gross-to-net dynamics during the first 3 years of a drug launch (Guth et al. 2024).

GROSS-TO-NET DEDUCTIONS BY EARLY LAUNCH YEARS, 2016 TO 2020

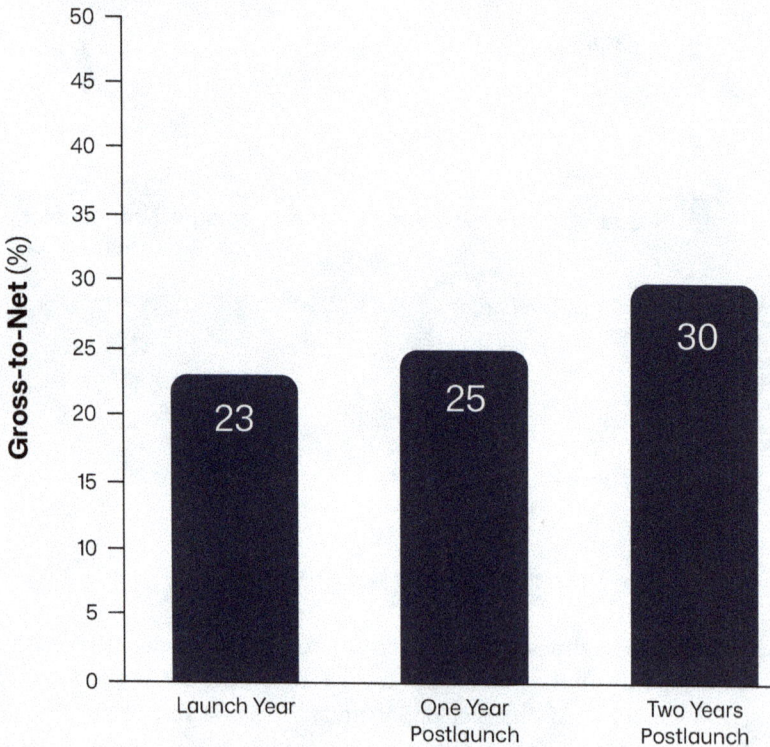

Source: Guth et al. 2024

Figure 42. Gross-to-net deductions for products in their first 3 years of commercialization.

As more patients are converted to traditional coverage for chronic conditions, the net price of each patient can improve moving forward. As product access and demand are established, manufacturers can become more aggressive to ensure volume is pulled through via broad, competitively favorable formulary placement when responding to insufficient pull-through demand, new competitive entrants, or changes in policy.

As discussed in Chapters 2 and 3, LOE and the entrance of generics or biosimilars becomes the ultimate threat to the gross-to-net as competition drives payers toward lower-cost options. Interestingly, research has shown that manufacturers who have heavy discounts in place may be protected due to "rebate walls." On the other hand, assets with modest discounts are more susceptible to market share erosion upon generic or biosimilar entry. Given this, oncology as previously discussed has the more modest gross-to-net reductions have been noted to be more susceptible to market share and net price erosion (Guth et al. 2024).

Key Considerations for Gross-to-Net Strategic Planning

Gross-to-net is a critical consideration for biotech and pharma companies. It reflects the discounts, rebates, and fees that reduce the list price, ultimately determining the actual net revenue received by the manufacturer. While we've explored various factors that impact gross-to-net, we'll leave you with several key questions to guide your own gross-to-net calculations for your asset:

1. What key factors should be assessed when developing a strategic plan to manage gross-to-net?

2. How does your asset's current market environment affect gross-to-net, and what competitive advantages or challenges should be considered?

3. What are the anticipated gross-to-net dynamics at launch, and how will they influence your product's life cycle?

4. What emerging payer trends or policy changes could affect your gross-to-net calculations?

APPENDIX 2
Incorporating Royalty, Upfront, and Milestone Payments in Valuation Models

· · · · · · · · · ·

> This appendix builds upon the licensing agreement concepts introduced in Chapter 5, providing a deeper examination of royalties, upfront payments, and milestone payments. This section will be helpful when incorporating licensing terms into your valuation model, understanding industry benchmarks for licensing deals, and analyzing how these payments affect cash flows for both licensors and licensees.

Licensing agreements regarding the authorized use of technology typically involve two key parties: the licensor and the licensee. Licensors are usually academic institutions, corporations, or individuals, with corporations being the predominant source, followed by universities and government agencies (Intellectual Property Research Associates 2022):

- ⅄ Corporations: 82%
- ⅄ Universities: 11%
- ⅄ Government Agencies: 3%
- ⅄ Hospitals: 2%
- ⅄ Individuals: 2%

Licensees are often large corporations or pharmaceutical companies looking to acquire rights to a technology for commercialization.

Valuing a licensing agreement requires analyzing the core deal terms: royalties, upfront payments, and milestone payments, which we examine in detail throughout this appendix. While beyond the scope of this guide, other important considerations in deal valuation include quantifying partner cost coverage (where the licensee may cover some or all development, regulatory, manufacturing, or other costs), cost and revenue synergies from overlapping operations or shared infrastructure, tax benefits from structural advantages, and deal financing and execution costs. These elements can materially impact valuations and should be considered alongside core licensing terms.

For comprehensive licensing and deal intelligence, Biotechgate (2025) provides a comprehensive database covering life sciences companies, clinical trials, and licensable assets across various sectors including biotech, pharma, medtech, and digital health sectors. It may be a useful resource when you're looking for the latest licensing benchmarks.

Royalty Rates

The majority of deals involve royalties as a percentage of net sales once the asset or candidate reaches the commercialization stage. Around 66% of deals include royalties (Neuendorf et al. 2023), which are either negotiated as a fixed percentage of all sales or structured as a tiered range of rates based on sales milestones. Another similar payment model is "profit-sharing," where instead of receiving a percentage of sales, the licensor and licensee split the actual profits generated by the product after deducting expenses from manufacturing, marketing, distribution, etc. While profit-sharing allows licensors to benefit more directly from a product's commercial success, it often leads to disputes over the amount of the profits shared between partners, which is why only a small fraction of deals adopt the "profit sharing" approach. Therefore, most agreements favor royalties over profit-sharing to ensure clearer financial terms between the licensor and licensee.

Figure 43 shows the distribution of royalty payments based on percentage of sales in deals from 1990 to 2022 (Intellectual Property Research Associates 2022). On average, around 50% of deals involve royalty rates of 6% or less, and around 90% of deals involve royalty rates of 20% or less.

DRUG LICENSING ROYALTY RATE DISTRIBUTION, 2022

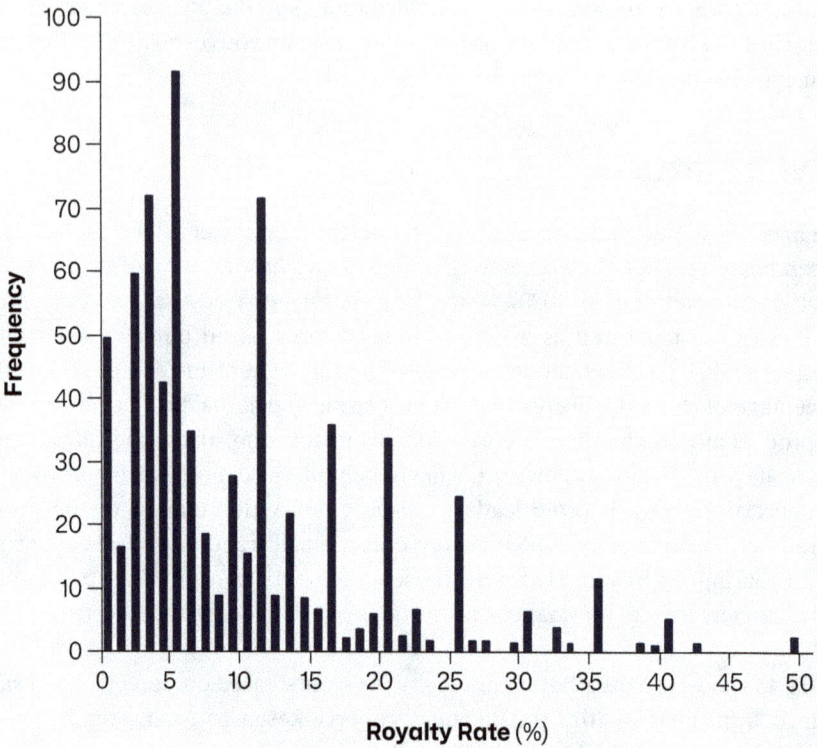

Source: Intellectual Property Research Associates 2022

Figure 43. Distribution of licensing royalty rates expressed as percentage of sales revenue.

Upfront Payments

Upfront payments are a common feature in biotech and pharma licensing deals due to their simplicity and straightforward terms. These payments provide immediate financial compensation to the licensor in exchange for granting rights to the licensee. Since the transaction is completed once the upfront payment is made, it eliminates uncertainty for both parties and secures the initial value of the deal.

A 2023 study analyzing 2,942 licensing deals globally from 2010 to 2023 found that around 87% of licensing agreements include an upfront payment to the licensor (Neuendorf et al. 2023). The average upfront payment across all clinical development phases was approximately $58.6M, with payments increasing as products advanced through clinical testing as seen in Figure 44 (Neuendorf et al. 2023). Phase 1 deals averaged $62.8M up front with total deal values reaching $450.3M. Phase 2 deals had slightly lower upfront payments averaging $61.4M with total deal values of $323.0M. Phase 3 deals, being closest to market, secured the highest upfront payments at $80.6M, though their total deal value averaged $311.7M. The total deal values encompassed upfront payments plus milestone payments tied to development achievements including IND filing, Phase 1 completion, Phase 2 results, Phase 3 success, regulatory approval, and sales targets.

AVERAGE UPFRONT AND TOTAL DEAL VALUES IN DRUG LICENCING, 2010 TO 2023

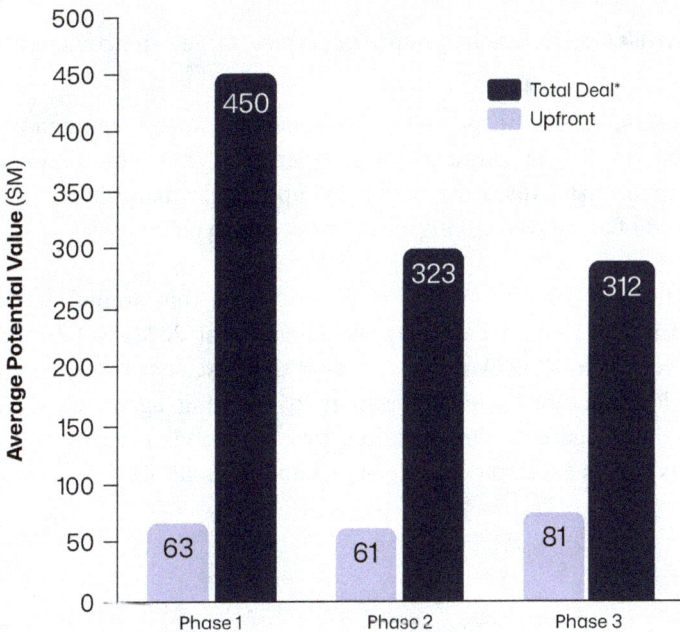

Total deal value encompasses upfront payments and milestone payments (IND, Phase 1, Phase 2, Phase 3, Approval, and sales)

Source: Neuendorf et al. 2023

Figure 44. Average upfront payments and total deal values for drug licensing agreements from 2010 to 2023.

Importantly, upfront payments are influenced by a drug's stage in clinical development. As a drug advances through trials, its perceived value slightly increases due to lower development risk and a clearer path to commercialization. This explains why upfront payments tend to be highest in Phase 3, as drugs in this stage have already demonstrated significant safety and efficacy with a clearer path towards commercialization, making them more attractive to potential licensing partners.

Milestone Payments

Milestone payments are made to the licensor when specific, pre-defined goals in the drug development or commercialization process are met. These milestones typically fall into two categories:

- **Development milestones:** Achieving key stages such as clinical trials (e.g., completing Phase 2 or securing FDA approval) or commercial launch.

- **Sales milestones:** Reaching a specific revenue target after commercialization.

In recent years, milestone payments have become more common compared to traditional royalty or profit-sharing structures. They now serve as a key component of licensing deals, second only to upfront payments in terms of timing and likelihood of being paid among all downstream payments.

The 2023 study of deals from 2010 to 2023 showed that around 86% of deals include a form of milestone payments (Neuendorf et al. 2023). Notably, milestone payments have been steadily increasing over the past decade, reflecting a shift toward performance-based compensation in licensing agreements. Based on the study's data, Figure 45 demonstrated two key trends in milestone payment distributions across development stages (Neuendorf et al. 2023).

AVERAGE PAYMENT VALUE BY DEVELOPMENT AND SALES MILESTONES, 2010 TO 2023

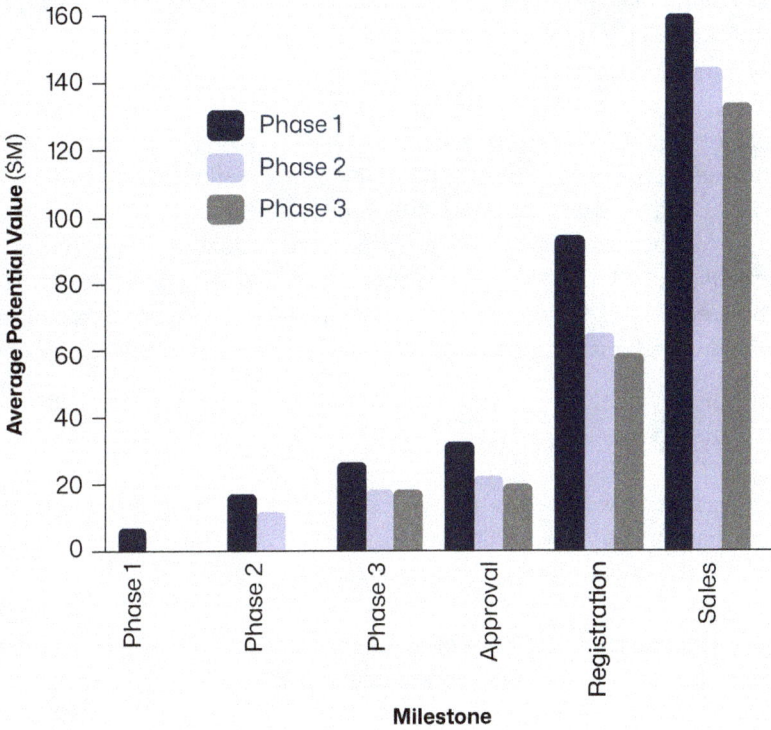

Source: Neuendorf et al. 2023

Figure 45. Average milestone payment values for development and sales achievements from 2010 to 2023.

First, total milestone payments tend to decline as products progress through clinical development, with average values of $387.5M in Phase 1, $261.6M in Phase 2, and $231.1M in Phase 3 (Neuendorf et al., 2023). This pattern likely exists because later-stage assets have fewer remaining development milestones to incentivize, and increased clinical certainty reduces the need for contingent payments. Both parties have clearer visibility into the product's value and sales potential based on available clinical studies.

Second, regardless of development stage, sales and registration milestones consistently represent the highest value components—accounting for approximately 67% of total milestone value across all phases. This payment structure demonstrates how risk is distributed in biotech partnerships, with the greatest rewards tied to commercial success rather than clinical development achievements.

In addition, average total milestone payments have increased substantially across all development stages based on a 2019 study that examined 218 partnerships between 2009 and 2013 and between 2014 and 2018 (Edwards 2019). The most dramatic growth occurred in early-stage deals, with discovery payments rising from approximately $30M to $70M and Phase 1 payments surging from $70M to $140M. This trend highlights the growing premium placed on securing "promising" assets earlier in development, particularly as competition for innovation intensifies.

Model Implementation

To keep things clear, we need to run two separate rNPV analyses: one for the licensor and one for the licensee. This is because their cash flows differ depending on their role in the deal (either as a source of revenue or cost).

When incorporating royalties, upfront payments, and milestone payments into a valuation model, these payments are categorized as either revenue (for the licensor) or expenses (for the licensee).

Example: Let's say a biotech product generates $40M in revenue in its launch year (2026) and $100M in revenue in the following year (2027). The licensing agreement, which was approved in 2024, includes the following terms:

- Royalty Rate: 5%
- Upfront Payment: $2M
- Development Milestone Payments:
  - New Drug Application Approval: $1M
  - Product Launch: $3M
- Sales Milestone Payments:
  - Revenue Target of $50M: $10M

Let's break down an example to demonstrate the impact of these payments on the cash flow of the Licensor and the Licensee, see Table 35 and Table 36, respectively. For simplicity purposes, we did not include other cash flow elements such as COGS, R&D, SG&A, etc.

LICENSOR PERSPECTIVE: CASH FLOW ANALYSIS BY YEAR

	2024	2025	2026	2027
	L-2	L-1	Launch (L)	L+1
	*Initial License Agreement	*Drug Approved		*Sales Surpass $50M
REVENUE (received from license agreement)	$2,000,000	$1,000,000	$5,000,000	$15,000,000
Royalty (5%)	-	-	$2,000,000	$5,000,000
Upfront	$2,000,000	-	-	-
Milestone	-	-	-	-
Development	-	$1,000,000	$3,000,000	-
Sales	-	-	-	$10,000,000

Table 35. Annual cash flow structure for licensor showing revenue streams and payment timing by year.

LICENSEE PERSPECTIVE: CASH FLOW ANALYSIS BY YEAR

	2024	2025	2026	2027
	L-2	L-1	Launch (L)	L+1
	*Initial License Agreement	*Drug Approved		*Sales Surpass $50M
REVENUE (generated from product sales)	-	-	$40,000,000	$100,000,000
COSTS	$2,000,000	$1,000,000	$5,000,000	$15,000,000
Royalty (5%)	-	-	$2,000,000	$5,000,000
Upfront	$2,000,000	-	-	-
Milestone	-	-	-	-
Development	-	$1,000,000	$3,000,000	-
Sales	-	-	-	$10,000,000

Table 36. *Annual cash flow structure for licensee showing revenue streams and payment timing by year.*

Notice the reciprocity between the licensor and licensee. The licensee generates revenue from the theoretical product starting in the launch year, which in turn funds royalty payments to the licensor. Additionally, the licensee makes milestone payments to the licensor upon achieving key development goals, such as application approval and sales targets. An upfront payment was also made at the time of the initial licensing agreement. The licensor's revenue then consists of these payments from the licensee.

These elements will be incorporated into our DCF analysis and discounted accordingly. The key takeaway is that while adding these payments is relatively straightforward from a modeling perspective, it's crucial to carefully consider their role as either a source of revenue or cost, depending on whether you are the licensor or licensee.

Key Considerations for Licensing Agreement Analysis

Licensing agreements reshape risk profiles and cash flow dynamics for licensor and licensee involved, impacting the overall valuation calculation. The strategic choice between licensing-out promising assets for immediate capital versus licensing-in technologies to accelerate development requires careful consideration of the risks involved, partnership strategies, market comparisons, and long-term accountability. While we've explored the various components and benchmarks for licensing deals, we'll leave you with several key questions to guide your own licensing strategy and valuation analysis:

1. What factors should drive your decision between being a licensor versus licensee, and how does this choice align with your company's strategic objectives and development capabilities?

2. How should current market benchmarks for royalty rates, upfront payments, and milestone structures influence your negotiation strategy based on your asset's development stage?

3. What is the optimal balance between development milestones and sales milestones in your payment structure to appropriately allocate risk and reward?

4. How will licensing arrangements affect your overall valuation model, and what sensitivity analyses should be conducted around key licensing assumptions in your DCF analysis?

APPENDIX 3
Implementing the Monte Carlo Approach in Valuation Models

. . . .

> This appendix expands on the rNPV concepts introduced in Chapter 6, offering an add-on framework for implementing Monte Carlo simulations in biotech valuation models. This section is helpful when you want to assess multiple inputs for a specific variable, usually the most uncertain or sensitive one, which may provide additional insights for your biotech valuation analysis.

Monte Carlo simulation is a powerful analytical technique that helps assess uncertainty in financial valuations. Rather than being a valuation method itself, it runs hundreds or thousands of scenarios to test how changes in key variables affect your final outcome. The process works by assigning probability distributions to uncertain inputs—such as growth rates, discount rates, or market conditions. Each simulation randomly selects values from these distributions and calculates a result. After running many trials, you get a range of possible outcomes that shows both the most likely results and the potential variability around them.

This approach is particularly valuable for identifying which assumptions have the greatest impact on your valuation. The simulation results can either serve as your final estimates or feed into additional analysis models, giving you a clearer picture of both expected returns and associated risks. The key advantage is moving beyond single-point estimates to understand the full spectrum of possibilities, making it easier to make informed decisions under uncertainty.

Monte Carlo simulations are particularly useful when valuing assets with complex, non-linear behaviors (Beaton 2019). Key examples include:

- **Path-Dependent Outcomes:** When value depends on the sequence of events over time, not just the final result.

  Example: A medication where the timing and order of doses affects effectiveness—taking the same total amount but in the wrong sequence or spacing could make the treatment ineffective.

- **Conditional Outcomes**: When an event only occurs if specific conditions are met first.

  Example: An insurance payout that only triggers if damage exceeds your $1,000 deductible—$999 in damage gets you nothing, but $1,001 gets you a check.

- **Non-Linear Outcomes**: When small changes in one factor can cause dramatically different results.

  Example: A few degrees difference in oven temperature might mean the difference between perfectly baked cookies and burnt ones.

While Monte Carlo simulations can be a powerful tool for analyzing uncertainty, they come with important limitations:

- **Garbage In, Garbage Out:** The results are only as good as your assumptions. If you guess wrong about how variables behave or relate to each other, your thousands of simulations will just give you precisely wrong answers.

- **False Precision:** Running 10,000 simulations might make your analysis look sophisticated, but it won't necessarily provide better insights than a well-thought-out scenario analysis with just three cases.

- **Overkill for Simple Problems:** If your outcomes change smoothly and proportionally with your inputs—without sudden jumps or critical tipping points—Monte Carlo will likely give you the same answer as just plugging in average values. Why run thousands of complex simulations when simple math gets you there faster?

Before applying Monte Carlo simulation, you need to clearly understand which variables drive your outcome and whether they create the kind of complexity that justifies this approach. Ask yourself: Does your valuation involve path-dependent sequences, conditional thresholds, or non-linear jumps that simpler methods can't capture accurately?

If your problem has these complexities, Monte Carlo could provide valuable insights into uncertainty that basic calculations miss. But if your outcomes change predictably with your inputs, you're likely better off with straightforward scenario analysis. The key is matching the tool to the problem—Monte Carlo shines when dealing with true complexity but becomes an expensive distraction when applied to straightforward situations.

Example Monte Carlo Application

To illustrate how this works in practice, let's apply Monte Carlo simulation to our biotech valuation example. Table 37 breaks down the key parameters in our rNPV model, organized by revenue and expenses—the same way you'd structure any valuation model to identify which variables need deeper analysis.

GENERAL CASH FLOW ELEMENTS FOR VALUATION: REVENUE AND EXPENSES

Revenue	Expenses
⋏ TAPP (Chapter 2)	⋏ COGS (Chapter 3)
⋏ Peak market share (Chapter 2)	⋏ R&D (Chapter 4)
⋏ Per-patient drug pricing (Chapter 2)	⋏ SG&A (Chapter 5)

Table 37. Overview of key cash flow components used in biotech company valuation models, including major revenue and expense categories.

- ⋏ **Revenue Parameters:** The TAPP and per-patient drug pricing can be estimated using market research, industry benchmarks, and competitor analysis.

- ⋏ **Expense Parameters:** COGS, R&D spending, and sales and administrative expenses can be benchmarked against industry standards and comparable company financial reports.

Among all these variables, peak market share stands out as the most uncertain parameter you'll encounter. It depends on multiple unpredictable factors: how quickly you reach market, whether you're first or fifth to launch, and how well your product differentiates itself. This uncertainty becomes even more pronounced for early-stage products that are still years away from launch.

Monte Carlo simulation lets us stress-test these peak market share projections systematically. The implementation is straightforward—we'll make minimal adjustments to our existing valuation model and run thousands of scenarios with different market share assumptions. We can then analyze the results using weighted averages of rNPV outcomes, calculate the probability of loss (percentage of scenarios where rNPV turns negative), visualize the results through histograms, and other statistical tools of your choosing.

Applying Monte Carlo to ONC-301

Let's return to our biotech example from Chapter 6. Our Excel model calculated an rNPV of approximately $125M for the biologic ONC-301, assuming a 30% peak market share. Since small changes in market share can dramatically alter the final valuation, we need to ask: How realistic is this 30% assumption, and what additional insights can we gain by testing different scenarios?

The commercial success of ONC-301 depends heavily on clinical trial results that will determine how well it differentiates from competitors on safety and efficacy. With the asset just entering Phase 1 trials, clinical development success remains highly uncertain and will certainly impact market share, making it the perfect variable for this Monte Carlo analysis example.

Here are the steps to implement the Monte Carlo method into our valuation model:

1. **Set Your Assumption:** Choose a mean peak market share and its standard deviation.

2. **Calculate Random Market Shares:** Create different market share scenarios based on probability distribution:

   a. **Generate Random Probabilities:** Determine the number of simulations and generate random values between 0–1 by using RAND() function in Excel.

   b. **Create Normal Distribution:** Use a normal distribution to generate different market share scenarios based on your peak market share, standard deviation, generated random values between 0–1 by using NORM.INV(generated_probability, mean_peak_market share, standard_deviation) function in Excel.

3. **Run Your Valuation Model:** Run rNPV analysis for each market share scenario (Use Data → What-If Analysis → Data Table in Excel).

4. Examine Your Results: Analyze three key outcome variables to determine the outcome of your simulated valuation model. These are defined as:

   a. **Number of Losses:** The total instances where rNPV scenarios resulted in negative values.

   b. **Probability of Loss:** The proportion of negative rNPV scenarios across all scenarios.

   c. **Average Outcome:** The weighted average of rNPV across all scenarios by using SUMPRODUCT(range of rNPV values, range of probability values)/SUM(range of probability values) in Excel.

In the "Monte Carlo" tab of the Excel sheet, you can select individual assets and set their peak market share and standard deviation. The tool then runs a Monte Carlo simulation using a normal distribution to generate multiple scenarios and calculate the corresponding rNPV outcomes. Using the data table containing the rNPV simulation values, a histogram can be plotted displaying the distribution of rNPV values grouped into buckets. The rNPV output is expected to follow the general shape depicted in the illustrative Monte Carlo simulation histogram in Figure 46.

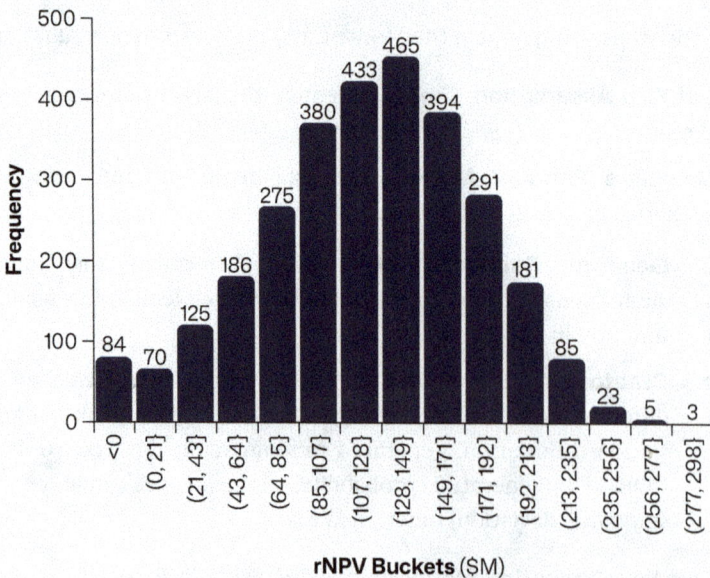

ILLUSTRATIVE MONTE CARLO SIMULATION: MARKET SHARE IMPACT ON VALUE DISTRIBUTION

Figure 46. Illustrative histogram showing distribution of rNPV values from Monte Carlo simulation with varying market share assumptions.

Using ONC-301 as an example with 3,000 simulations, a mean peak market share of 30%, and a standard deviation of 15%, the Monte Carlo analysis can provide additional insights into the risk of investment. The simulation showed 84 possible negative outcomes with negative rNPV, translating to a roughly 3% probability of loss that the project becomes financially inviable, while generating a weighted average rNPV of +$152M.

The results help highlight both the potential and the risks of the investment. While the majority of scenarios support a financially viable project, approximately 3% of potential outcomes could result in negative returns. Notably, the weighted average rNPV of +$152M represents a 22% increase over the baseline single-point estimate of +$125M, which is expected since higher market share scenarios lead to greater growth in the rNPV output. However, even a 3% probability of loss represents significant capital at risk, particularly given the binary nature of clinical trial outcomes that could dramatically shift these projections as ONC-301 progresses through its development phases.

In summary, the Monte Carlo method provides an additional tool for biotech valuation to help prioritize drug candidates that have a low probability of making a loss and high weighted average rNPV across many scenarios. Such prioritization is helpful for gaining investors' interest in your biotech asset or drug candidate.

Key Considerations for Monte Carlo Simulation Implementation

Monte Carlo simulation offers a dynamic approach to uncertainty analysis in biotech valuation, particularly valuable for assets with path-dependent, conditional, or non-linear outcomes. However, its complexity and computational demands mean it should be applied judiciously—only when simpler analytical methods cannot adequately capture the underlying risk dynamics of your valuation model. While we've explored the implementation framework and analytical benefits of Monte Carlo methods, we'll leave you with several key questions to guide your decision on whether and how to incorporate this technique:

1. Does your biotech asset exhibit path-dependent, conditional, or non-linear characteristics that cannot be accurately captured through traditional scenario analysis or sensitivity testing?

2. Which variable in your valuation model carries the highest uncertainty and greatest impact on outcomes, and would Monte Carlo simulation provide meaningful insights beyond standard deviation analysis?

3. How reliable are your input distribution assumptions and correlations, and do you have sufficient data quality to justify the added complexity of Monte Carlo modeling?

4. What specific decision-making value will the probability of loss calculations and weighted average outcomes provide to stakeholders that simpler valuation approaches cannot deliver?

APPENDIX 4
Alternative Approaches to Biotech Valuation

This appendix introduces alternative approaches to biotech asset valuation beyond the rNPV method covered in Chapter 6. These techniques are particularly valuable when working with VC investors or when quantifying the value of flexibility in business decision-making. Understanding these complementary methodologies enables additional asset assessment across different investment contexts and negotiations.

Comparable Multiples and Venture Capital (VC) Valuation Method

The Comparable Multiples method provides a rapid valuation approach by applying financial ratios from similar companies to estimate a target company's value. This method relies on market-based multiples such as price-to-earnings (P/E), price-to-sales (P/S), or enterprise value-to-revenue (EV/Revenue) ratios derived from comparable public companies or recent transactions. This approach is most effective for companies with established revenue streams, as the underlying multiples require meaningful financial metrics to produce reliable valuations. The method assumes that similar companies should trade at similar valuation multiples, adjusted for differences in growth, profitability, and risk profiles.

The VC Valuation Method combines the Comparable Multiples Method with DCF principles to determine the appropriate equity stake for a given investment amount. This hybrid approach calculates the ownership percentage required to achieve target returns by considering expected exit valuation based on comparable multiples, time horizon to exit, required rate of return (discount rate), and risk assessment.

The VC Valuation Method involves these key steps:

1. **Exit Timing:** The VC firm first determines an "exit date," which is the point at which they intend to sell their stake. This could be through an acquisition or an initial public offering (IPO).

2. **Estimating Enterprise Value:** Next, the VC estimates the company's enterprise value (EV), the total value of the business independent of how it's financed, at the exit date using the Comparative Multiples method by using ratios or multiples derived from public companies, market trends, and current prices. For instance, as of January 2025, biotech companies were typically valued with EV/sales and EV/EBITDA multiples of 6x and 15x, respectively (Damodaran 2025a; Damodaran 2025b). The process involves:

   a. Projecting the company's financial metrics (revenue, EBITDA, etc.) at the anticipated exit date.

   b. Identifying comparable publicly traded companies or past transactions.

   c. Analyzing price multiples such as the price-to-earnings ratio (P/E) or earnings/ EBITDA/sales multiples.

   d. Applying these multiples to the target asset's financial metrics.

   e. Adjusting for any differences between the comparable companies and the subject asset.

3. **Calculating Company's Present Value using DCF analysis:** The VC calculates the company's NPV by discounting the estimated future enterprise value to present value using a target rate of return as the discount rate over the projected years until exit. As a reminder, when using NPV without risk adjustment, this rate reflects the total perceived investment risk, including both drug-related and non-drug-related uncertainties. Early-stage startups typically command discount rates of 50–70% at the seed stage, decreasing to 30–60% in later funding rounds as companies mature, as shown in Figure 47 (Aggarwal and Bahl 2021).

4. **Estimating Percentage Ownership:** VCs calculate their equity stake using the post-money valuation formula: Investment Amount ÷ Post-Money Valuation = Ownership Percentage. This ensures an accurate representation of their actual ownership after the funding round closes.

> **Quick tip:** One important concept is defining "pre-money" and "post-money" valuations, which influence the calculated equity stake. Pre-money valuation is the company's value before receiving new investment, while post-money valuation is the company's value after the investment is added. For example, if a company has a $15M pre-money valuation and an investor provides $5M, the post-money valuation becomes $20M ($15M + $5M), giving the investor 25% ownership of the company ($5M divided by $20M). Investors use post-money valuation because they want to know their actual equity stake after the company receives funding—using pre-money would overestimate ownership at 33% ($5M divided by $15M).

VENTURE CAPITAL NPV FRAMEWORK: DISCOUNT RATES AND RETURN MULTIPLES

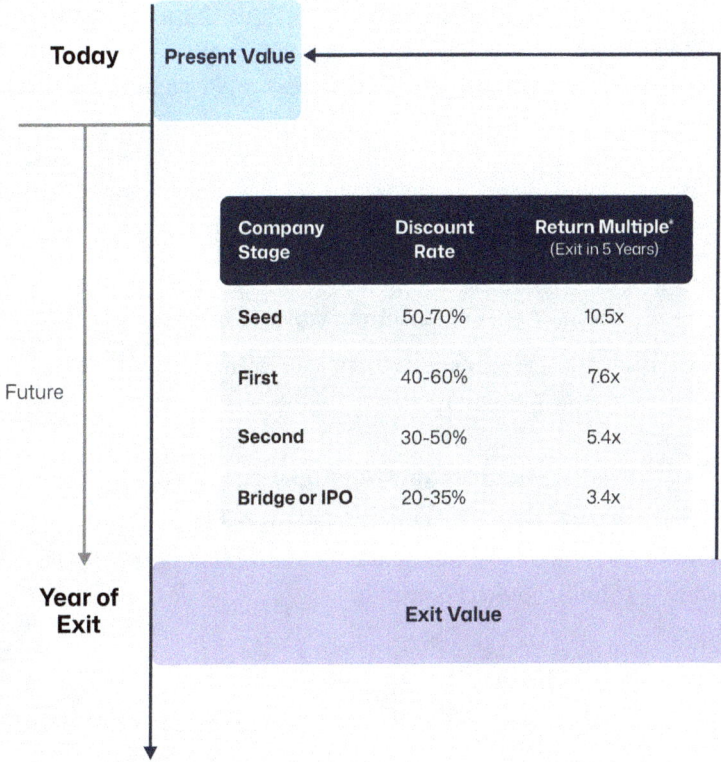

Today **Present Value**

Future

Company Stage	Discount Rate	Return Multiple* (Exit in 5 Years)
Seed	50-70%	10.5x
First	40-60%	7.6x
Second	30-50%	5.4x
Bridge or IPO	20-35%	3.4x

Year of Exit

Exit Value

Calculation: (1 + discount rate)^years to exit. Example: $1M seed investment at 60% discount rate requires $10.5M company valuation at 5-year exit: (1+60%)^5 = 10.5x return multiple

Note: Bridge stage financing supports companies in pilot plant construction, production design, testing, and pre-IPO preparation activities.

Source: Aggarwal and Bahl 2021

Figure 47. NPV framework for VCs showing discount rates and expected return multiples by company development stage over a 5-year investment horizon. Discount rates reflect total investment risk (drug development and non-drug factors) using unadjusted NPV.

Note that this method does not account for R&D costs and probabilities of success of each clinical phase. As a drug progresses through clinical phases, earlier R&D costs lose relevance, and the drug's value is based on expected future cash flows and remaining costs. In contrast, the VC method combines future revenues and all development costs into a single valuation multiple at the exit date, with risks handled by a high hurdle rate. This means the rNPV method captures the drop in R&D costs as development moves forward, while the VC method doesn't account for this change.

In addition, the Comparable Multiples and VC Valuation Method comes with certain challenges, especially when applied to novel biotech assets:

01
Finding Applicable Comparables is Difficult
Identical assets are rare, and data from similar deals can quickly become outdated.

02
Market Multiples May Not Apply
Publicly traded biotech and pharma companies may trade at premiums or discounts that don't reflect the true value of private biotech companies.

03
Emphasis on Historical Performance
Biotech multiples often reflect past performance rather than future potential, which can result in inflated valuations for companies with strong historical results but poor future prospects.

04
Revenue-Based Multiples Don't Work for Pre-Revenue Companies
For early-stage, pre-revenue biotech companies, using revenue-based multiples can undervalue the business since these metrics don't account for the potential of early-stage technologies.

05
Opaque Deal Terms in Licensing
In licensing deals, only the upfront payments or headline values are typically disclosed, leaving out key details that would provide a fuller picture of the deal's value.

In such cases, the rNPV and SOTP method is often a more reliable and suitable approach to accurately value biotech assets. The Comparable Multiples and VC Valuation Method can be used as a sanity check for your calculations.

Real Options Valuation Method

The DCF method and its variants, like the rNPV and VC methods, are popular for valuing biotech assets because they're straightforward to use and understand. But these methods often take a "set it and forget it" approach, assuming the company will stick to a fixed plan no matter what happens. In the real world, though, things can change as new side effects might emerge, or competition in the market could change. That's when management can step in and adjust the strategy, re-evaluating whether to keep moving forward or abandon a project if further R&D investment doesn't make sense anymore.

This is where the Real Options Valuation method shines. Unlike traditional methods, real options method takes into account the flexibility management has to change course based on new information. In pharma R&D, this flexibility is crucial, as companies constantly decide whether to move forward, pause, or scrap a project based on how it's progressing. Real options models focus on:

- **Strategic Flexibility:** The ability to expand, pause, or terminate projects at key milestones.
- **Uncertainty Management:** Accounting for market fluctuations, clinical outcomes, and regulatory challenges.
- **Decision-Driven Valuation:** Assessing how active decision-making impacts the asset's overall value.

Here's are the notable steps in the Real Options Method:

1. **Determine the Asset's Current Value**
   a. The asset's present value is based on projected future cash flows if it successfully reaches commercialization.

2. **Model Future Value Scenarios with a Binomial Tree**
   a. Since biotech assets have high uncertainty, their future value fluctuates based on various factors (e.g., clinical trial success, competitive landscape).

b. A binomial tree maps out potential value trajectories at each stage, showing possible increases or decreases over time.

c. Higher volatility leads to a wider range of potential outcomes, increasing the importance of strategic flexibility.

3. **Factor in Key Investment Decisions**

a. At each development stage (e.g., clinical trials, regulatory submission), companies must decide whether to proceed, delay, or abandon further investment.

b. By working backward from the final stage, we calculate the value of maintaining flexibility, incorporating costs and risks associated with each decision point.

The final output, the "real option value", reflects the present worth of an asset today, incorporating the ability of a company to adapt. Unlike rNPV and comparative valuation methods, which assume a fixed path, real options valuation captures the managerial flexibility needed to navigate a biotech company's uncertainties. By quantifying the value of strategic decision-making, it provides a slightly more involved but more realistic and dynamic approach to a biotech valuation.

Key Considerations for Alternative Valuation Method Selection

Alternative biotech valuation techniques provide complementary perspectives to traditional rNPV analysis, each serving distinct purposes across different investment contexts and asset characteristics. The Comparable Multiples and VC methods offer rapid assessment tools particularly valuable for later-stage assets and investor negotiations, while Real Options valuation captures the strategic flexibility inherent in biotech R&D decision-making. While we've explored these methodologies and their applications, we'll leave you with several key questions to guide your selection of the most appropriate valuation approach:

1. Which valuation method best aligns with your asset's development stage, revenue profile, and the specific decision-making context you're addressing?

2. Do you have access to reliable comparable company data and market multiples that accurately reflect your asset's unique characteristics and market position?

3. How significant is managerial flexibility and strategic decision-making optionality in your asset's development pathway, and would Real Options valuation provide meaningful insights beyond traditional DCF methods?

4. What combination of valuation approaches would provide the most comprehensive assessment for stakeholder presentations, and how should these methods be weighted in your final investment or strategic recommendations?

APPENDIX 5
Technology Licensing in Academic Institutions

This appendix expands on licensing concepts from Chapter 5, providing a detailed examination of university technology transfer processes. It outlines the general steps of the technology transfer process for inventors used by leading academic institutions and explains key licensing terms that entrepreneurs or startups encounter, including equity stakes, royalty rates, milestone payments, and performance obligations. Use this section as a reference when evaluating academic licensing opportunities, negotiating university partnerships, and assessing both the commercial potential and structural challenges of university-developed technologies.

The Promise and Reality of Technology Transfer

While these institutional frameworks provide the foundation for commercialization, the actual outcomes reveal a complex picture of both remarkable successes and sobering realities. Among the top 10 U.S. research institutions, roughly 90% of licensing income comes from life-science intellectual property, including patents and drug discoveries (Prasad 2025). A handful of blockbuster biotech patents typically account for the lion's share of licensing revenue, creating impressive success stories like Columbia University, which generated $678M in licensing revenue from 2012 to 2015, largely from medical research patents, while spawning 74 startup companies (Department for Science, Innovation & Technology 2023).

However, such cases represent the exception rather than the rule. The broader reality reveals significant challenges: nearly 60% of university licenses generate no licensing revenue at all, and only about 1% ever earn over $1 million for the university (Prasad 2025). This economic reality forces most Technology Transfer Offices (TTOs)—the university departments responsible for commercializing academic research and managing licensing agreements—to operate at a loss, leaving directors to navigate between their stated mission of maximizing societal impact and institutional expectations to function as revenue-generating centers.

The Standard Technology Transfer Process

Despite these challenges, universities have provided a standardized pathway for inventors to pursue commercialization success. Using Stanford's TTO as a framework (Stanford Office of Technology Licensing 2025), the typical technology transfer process includes nine key phases:

1. **Disclosure**
   a. Inventors submit a disclosure form recording invention details, inventors, sponsors, and public disclosures.
   b. Primary investigator authorization is required for lab-specific inventions.

2. **Manager Assignment**
   a. Each disclosure receives a docket number and assignment to a licensing manager.

3. **Evaluation**
   a. The licensing manager and inventors assess patentability, novelty, and market potential.
   b. Inventors contribute to prior art searches and licensing strategy decisions (exclusive versus non-exclusive).
   c. Strategy depends on invention nature and expected development costs.

4. **Patent Application**
   a. Not all inventions are patented—a decision is made based on commercial and legal potential.
   b. When pursued, inventors collaborate with patent attorneys on drafting and revision.
   c. Inventor input is critical for meaningful patent protection.

5. **Marketing**

   a. TTOs market inventions using non-confidential abstracts (with inventor input).

   b. Inventors help identify potential licensees and answer technical questions under a non-disclosure agreement (NDA).

6. **Negotiations**

   a. The licensing manager leads negotiations with inventor consultation on company fit and conflict assessment.

   b. Only approximately 20–25% of disclosures result in licenses.

7. **Monitoring Progress**

   a. Post-licensing, TTOs monitor company development and financial progress through regular reports.

8. **Patent Income Sharing (Royalty and Equity)**

   a. After expenses, royalties and equity proceeds are shared annually.

   b. Multiple inventors typically split proceeds equally unless otherwise agreed.

9. **License Amendments**

   a. Licenses can be updated over time as circumstances change.

   b. Licensing managers negotiate amendments as needed.

> **Note:** Policies vary significantly between universities. Consult your institution's TTO for the most current information.

After the inventor and TTO finalize their agreement, the technology becomes available for licensing to companies and entrepreneurs and can start generating revenue.

Revenue Sharing Between Inventors and Universities

Many U.S. universities distribute patent income with inventors receiving approximately one-third, while the remaining two-thirds is allocated among various university entities including departments, schools, and central operations, though specific allocations vary significantly by institution. For example, Stanford's patent income distribution follows this general framework after first deducting 15% from gross patent income for TTO administrative overhead and any amounts owed to other institutions or entities. Of the remaining distributable income, inventors receive one-third (33.3%) while the university allocates the remaining two-thirds between departments (24.7%), schools (21.0%), and research operations (21.0%) for income under $3M. For patents generating over $3M, inventors maintain their one-third (33.3%) share, but Stanford reallocates its portion to increase funding for schools and research operations (26.0% each) while reducing departmental allocation (14.7%). This example structure reflects the partnership where inventors provide breakthrough discoveries while universities contribute the infrastructure, legal expertise, and risk capital necessary to transform innovation from the bench into market success.

Understanding University Licensing Terms

While revenue sharing defines the long-term partnership between inventors and universities, entrepreneurs must first navigate the licensing terms that govern access to these technologies. For startup founders and entrepreneurs seeking to utilize university technologies, understanding standard licensing terms and their underlying rationale becomes essential for successful negotiations. Universities typically structure deals around several key components that reflect both their need to recoup research investments and their mission to maximize societal impact. While specific terms vary significantly between institutions and technologies, understanding this common framework helps entrepreneurs anticipate what universities seek during negotiations.

Important terms that startups will encounter include (Prasad 2025):

- **Field of Use and Exclusivity:** University licenses can be exclusive for specific fields/markets or non-exclusive, with exclusive rights being crucial for attracting investors who need assurance competitors won't receive the same technology.

- **Upfront Fees:** License fees range from $1,000–$10,000 for startups to $10,000–$250,000 for established companies. Many institutions waive or defer startup fees until capital is raised.

- **Equity Stakes:** Universities increasingly request equity stakes (typically 5% post-Series A, diluting to 1–2% at exit) alongside or instead of cash fees to align incentives with startup success. Universities may seek antidilution protection, though founders often resist since it can complicate the cap table, which is a spreadsheet of information regarding a company's percentages of ownership, equity dilution, and value of equity in each round of investment.

- **Royalties on Sales:** Common startup royalty rates range from 1–5% of net sales, with 2% being the most frequent. Academic biotech licenses average lower rates (~3%) than company-to-company deals (~8%) since university inventions are early-stage.

- **Milestone Payments:** Common in biotech, these lump sum payments trigger at specific milestones like clinical trials or approval. Most institutions will renegotiate milestone timing for good faith progress.

- **Patent Cost Reimbursement:** Startups must cover ongoing patent prosecution and maintenance costs, often reimbursing past university expenses over several years.

- **Performance Obligations (Diligence):** Licenses require active commercialization with development milestones by specific dates. Failure to meet obligations can result in license termination or conversion to non-exclusive status.

- **Sublicensing and Subcontracting:** Startups may sublicense technology to partners, with universities typically requiring 10–50% of sublicense income sharing. Agreements distinguish between sublicenses and routine supplier relationships to avoid fees on normal operations.

- **Know-How and Confidential Information:** Universities often bundle tacit knowledge, methods, and lab data with patents, with royalties continuing even if patents don't issue. This prevents companies from avoiding payments by relying solely on unpatented knowledge.

- **Assignment and Exit Terms:** Agreements specify terms for startup acquisitions, typically allowing assignment to acquirers with university approval. Some universities negotiate "success fees" or increased payouts for acquisitions within certain timeframes.

- **Governance Rights:** Universities occasionally request board observer seats, more common with larger equity stakes. Actual voting board seats are rare in these arrangements.

To illustrate how these terms play out in practice, here's how several leading research universities approach licensing and startup partnerships. The following comparison in Table 38 draws from publicly available policy documents (Prasad 2025):

UNIVERSITY TECHNOLOGY LICENSING: IP AND STARTUP POLICIES

Policy Area	Stanford	MIT	University of California (UC)	Harvard
IP Ownership Model	University-owned, inventor-participation	University-owned	University-owned	University-owned
Typical Equity Stake	1-5%	1-5% (U.S. average)	Variable	1-5% (U.S. average)
Royalty Rates	Varies by sector	~2% (common rate)	Percentage of net sales	~2% (common rate)
Upfront Licensing Fees	Requires some cash upfront	Case-dependent	Upfront fees required	Case-dependent
Additional Comments	Deal flexibility of royalties, equity, and fees	Possible to have one invention with multiple licensees	Patent cost reimbursement, annual maintenance fees, sublicensing fees	Global access considerations for healthcare technologies

Source: Prasad 2025

Table 38. Comparative overview of IP policies, startup terms, and technology licensing frameworks across major U.S. research universities.

Note: These policies represent general frameworks and may vary by specific technology, market sector, and individual negotiations. Founders should consult directly with each institution's Technology Transfer Office for current terms.

Key Considerations for Strategic University Technology Licensing

University technology transfer represents a complex ecosystem where academic innovation meets commercial reality, requiring careful navigation of institutional frameworks, licensing terms, and strategic partnerships. The structured processes employed by leading TTOs provide clear pathways for commercialization, yet the stark statistics—with only 1% of licenses generating over $1 million and most TTOs operating at a loss—underscore the inherent challenges in translating laboratory discoveries into market successes.

While standardized licensing frameworks offer predictability, the negotiable nature of most terms creates opportunities for entrepreneurs to structure deals that align with their venture's specific needs and growth trajectory. As you consider engaging with university technology transfer, several critical questions should guide your strategic approach:

1. Does your venture's development timeline and capital structure align with the university's performance obligations and milestone requirements, and can you realistically meet the diligence standards within the specified timeframes?

2. How will the proposed licensing terms—particularly equity stakes, royalty rates, and exclusivity arrangements—impact your ability to attract subsequent investment rounds and maintain attractive returns for future investors?

3. Which university partners offer not just the technology you need, but also the institutional flexibility, industry connections, and ongoing research collaboration that could accelerate your commercialization efforts?

4. What combination of upfront fees, equity participation, and performance-based payments would optimize your relationship with the university while preserving sufficient value capture for your venture's long-term success?

Resources

T his section provides essential tools and reference materials designed to support your practical application of the valuation methodologies presented throughout this book. These resources include extra reference data, downloadable models, and systematic checklists that will help you implement rigorous biotech valuation analysis in real-world scenarios. Use these materials to enhance your understanding and streamline your valuation workflow.

Resource 1

Comprehensive Excel Model: Biotech Valuation Toolkit

A comprehensive Excel valuation model that incorporates all the methodologies and frameworks taught throughout this book is available for download from our website. This interactive tool will guide you through the valuation process step-by-step, making it easier to apply the concepts you've learned to real-world scenarios.

You can access this model by visiting agmigroup.com.

Navigate to the "Resources" section, where you'll find detailed information about this book and the downloadable Excel file. The model is specifically designed to complement the example valuation case presented in Chapter 6 and includes built-in assumption tables, patient population calculators, DCF analyses, and optional Monte Carlo simulation. This comprehensive toolkit will serve as a valuable template for conducting your own biotech company valuations.

Clinical Trial Cost Calculation: Per-Patient Costs and Enrollment Numbers

Table 39 presents per-patient costs and patient enrollment figures from the same data presented in Chapter 4, organized by therapeutic area and development phase (Sertkaya 2024). Use it to estimate clinical trial costs for your assets by selecting relevant per-patient costs and expected number of enrolled patients.

PER-PATIENT COSTS AND ENROLLMENT NUMBERS BY THERAPEUTIC AREA AND DEVELOPMENT PHASE

Therapeutic Area	Per-Patient Cost, 2025 $K (Enrolled Patients Per Trial)			
	Phase 1	Phase 2	Phase 3	Phase 4
Anti-infective	24.7 (69)	75.5 (243)	38.2 (575)	17.6 (1430)
Cardiovascular	75.7 (42)	52.6 (189)	42.1 (1151)	43.2 (508)
Central nervous system	111.2 (44)	62.1 (243)	50.4 (529)	44.5 (356)
Dermatology	45.1 (106)	84.9 (225)	61.9 (568)	42.1 (850)
Endocrine	108.8 (38)	65.6 (225)	62.1 (414)	72.3 (482)
Gastrointestinal	78.7 (38)	81.0 (292)	60.7 (496)	67.1 (1344)
Genitourinary system	68.4 (50)	58.3 (323)	49.6 (546)	21.3 (410)

Hematology	444.7 (31)	128.0 (123)	150.8 (233)	53.4 (411)
Oncology	131.6 (58)	100.3 (137)	118.6 (293)	29.9 (261)
Respiratory system	56.4 (49)	55.5 (203)	59.5 (516)	24.2 (1159)
Ophthalmology	64.9 (121)	61.7 (299)	101.8 (876)	30.6 (413)
Pain and anesthesia	115.0 (36)	98.9 (270)	77.3 (1209)	52.9 (280)
Immuno-modulation	80.8 (55)	61.0 (323)	69.9 (309)	38.5 (383)
All indications	**103.5 (51)**	**74.6 (235)**	**67.7 (630)**	**44.8 (708)**

Source: Sertkaya et al. 2024

Table 39. Breakdown of clinical trial per-patient costs and patient enrollment numbers across therapeutic areas and development phases.

Total Addressable Patient Population Estimation Checklist

TAPP estimation is the fundamental first component of market size analysis in biotech valuation. This systematic approach converts disease epidemiology into realistic patient numbers by addressing incidence/prevalence, patient segmentation, diagnosis rates, treatment eligibility, and medication adherence—ensuring your revenue projections avoid common overestimation pitfalls.

This checklist is designed to walk you through the key steps involved in building a TAPP estimate—from identifying the disease burden to adjusting for real-world factors like diagnosis, treatment eligibility, and patient adherence.

You should use this checklist whenever you are modeling market opportunity for an early-stage product. It helps ensure that your projections are grounded in data, reflect clinical realities, and avoid common overestimation pitfalls.

Understand the Disease Burden

- ☐ Have I reviewed both incidence and prevalence data from trusted sources (e.g., government health data, peer-reviewed medical journals, reputable industry benchmarks)?

- ☐ Have I consulted with medical experts (healthcare providers, clinical researchers, or epidemiologists) to determine which metric is more relevant for my specific therapeutic context?

- ☐ Have I evaluated whether the disease characteristics align with incidence or prevalence?

- ☐ Have I considered disease-specific exceptions where standard rules don't apply (e.g., DVT requiring long-term treatment, gout's acute flare-ups versus chronic management)?

Subdivide the Patient Population

- ☐ Have I adjusted for clinical eligibility (e.g., disease severity, stage, type, diagnostic imaging/tests, specific biomarkers, risk stratification)?

- ☐ Have I accounted for key demographic factors (age, gender, race, ethnicity) and their impact on treatment eligibility and outcomes?

- ☐ Have I considered comorbidities and contraindications that would exclude patients from treatment or require different therapeutic approaches?

- ☐ Have I factored in physician clinical judgment variations and consulted medical experts where established subdivision data may be limited?

- ☐ Have I clearly documented all subdivision assumptions with supporting rationale and data sources for each patient segment?

Estimate the Diagnosis Rate

- ☐ Have I assessed current diagnostic capabilities including availability of screening tests, diagnostic technology advancements, and sensitivity/specificity improvements?

- ☐ Have I considered reimbursement policies and insurance coverage for diagnostic tests and their impact on screening frequency and diagnosis rates?

- ☐ Have I considered healthcare access and infrastructure variations across different regions, healthcare systems, and socioeconomic populations?

- ☐ Have I evaluated the impact of asymptomatic versus symptomatic presentations on diagnosis rates, including screening program effectiveness for specific conditions?

- ☐ Have I accounted for evolving clinical definitions and diagnostic criteria that may increase or decrease diagnosis rates over time (e.g., multiple myeloma)?

Estimate the Treatment and Access Rate

- ☐ Have I evaluated current treatment standards and physician practice variations across different providers?

- ☐ Have I considered patient treatment preferences including tolerance for side effects, route of administration, and "wait and see" approaches?

- ☐ Have I assessed non-medical alternatives and their impact on treatment rates (e.g., lifestyle changes, surgical options, alternative therapies)?

- ☐ Have I analyzed access barriers including insurance coverage, geographic disparities, distribution constraints, and out-of-pocket costs?

- ☐ Have I accounted for treatment positioning (first-line versus last-resort) and estimated the percentage of patients who would exhaust other options first?

Adjust for Adherence and Compliance

- ☐ Have I broken down adherence into three key components: filling the first prescription, refilling prescriptions, and taking doses as prescribed?

- ☐ Have I adjusted adherence rates based on treatment setting: outpatient acute care, outpatient chronic care, and inpatient/physician-administered care?

- ☐ Have I considered disease-specific factors including symptom presence (asymptomatic versus severe conditions), disease severity, and patient perception of medication effectiveness?

- ☐ Have I factored in drug formulation and other factors including route of administration, dosing frequency, combination therapies, and branded/generic status?

Avoid Common Pitfalls

- ☐ Have I correctly differentiated and applied incidence versus prevalence based on the specific disease characteristics and treatment context rather than broad generalizations?

- ☐ Have I used high-quality, representative data sources and considered modeling different scenarios for rare or evolving conditions with limited or variable data?

- ☐ Have I ensured demographic representativeness of the target population including age, gender, education level, race, and ethnicity?

- ☐ Have I documented all key assumptions, limitations, and uncertainty ranges with supporting rationale?

Resource 4

Market Share Estimation Checklist

Market share projection is the critical second component of market size analysis in biotech valuation. This systematic approach transforms your addressable patient population into realistic revenue forecasts by analyzing competitive positioning, market entry timing, and product differentiation—ensuring your projections reflect marketplace realities rather than overly optimistic commercial assumptions.

This checklist is designed to walk you through the key steps involved in developing a credible market share estimate—from mapping the competitive landscape to evaluating your product's distinctive advantages and accounting for order-of-entry effects.

You should use this checklist whenever you are modeling market opportunity for an early-stage product. It helps ensure that your projections are grounded in competitive intelligence, reflect market dynamics, and avoid common overestimation pitfalls that can undermine the credibility of your valuation model.

Competitive Landscape Analysis

- ☐ Have I identified all current market competitors using industry reports, online databases, and patent records to map companies working on similar therapies?

- ☐ Have I researched late-stage pipeline competitors (Phase 2/3) that are likely to launch before my product enters the market?

- ☐ Have I appropriately categorized competitors by development risk including late-stage competitors in my model while excluding very early-stage or high-risk competitors?

- ☐ Have I analyzed the current "gold standard" or market leader to understand what drives physician treatment decisions and how current standards meet physician and patient needs?

- ☐ Have I evaluated each competitor's differentiating factors including clinical data strength, side effect profiles, dosing convenience, and care setting requirements?

Market Entry Timing Assessment

- ☐ Have I determined my order of market entry and expected launch timing relative to first-in-class and other identified competitors?

- ☐ If not first-in-class, have I calculated the expected time gap between the first entrant and my product and made market share adjustments accordingly?

- ☐ Have I factored in market share impact on first-mover advantage understanding that advantages are greater in smaller and less differentiated markets?

- ☐ Have I considered additional timing factors, including company size advantages (e.g., large pharma versus smaller biotech companies) and unmet clinical needs (e.g., eligibility for accelerated approval pathways)?

Product Differentiation Evaluation

- ☐ Have I identified the clinical and convenience advantages my product offers to determine whether it should be classified as well differentiated or undifferentiated compared to current and future competitors?

- ☐ Have I evaluated clinical factors including efficacy, safety profile, side effects, and any breakthrough therapy designations?

- ☐ Have I assessed convenience factors such as route of administration, dosing frequency, combination therapy, and other patient compliance advantages?

☐ Have I identified targeted therapeutic advantages in specific patient subgroups with unmet needs, breadth of indications, or ability to address treatment failures from existing therapies?

Special Considerations

☐ Have I applied appropriate market share ranges using conservative estimates for undifferentiated products versus differentiated products?

☐ Have I considered therapeutic class dynamics and whether the class welcomes later entrants?

☐ Have I considered multiple competitor scenarios including early, expected, and delayed competitive entry timelines to stress-test market share assumptions?

☐ Have I evaluated commercial execution advantages such as sales force strength, marketing strategy, and ability to carve out market niches?

☐ Have I assessed pricing strategy implications for market share, including competitive pricing pressures and premium pricing sustainability for differentiated products?

☐ If using physician survey data to estimate uptake, have I adjusted for physician overestimation bias and considered payer restrictions, patient fill rates, and commercial reach limitations?

Validation and Reality Check

☐ Have I benchmarked against analogous product launches with similar competitive dynamics, differentiation levels, and market characteristics to validate my analysis?

☐ Have I used questioning approaches that force physicians to consider tradeoffs between treatment options within specific patient segments rather than estimating uptake in isolation?

☐ Have I stress-tested my assumptions against multiple competitor entry scenarios?

☐ Have I documented all key assumptions and limitations with supporting rationale, data sources, and sensitivity analyses to facilitate stakeholder discussions?

Glossary

340B Program   A federal program requiring manufacturers to provide significant drug discounts to eligible hospitals and clinics serving low-income and uninsured patients.

Accelerated Approval   FDA pathway that allows drugs for serious conditions to be approved based on surrogate endpoints rather than waiting for long-term outcomes.

Adherence Rate   The percentage of patients who consistently take their prescribed medications as directed over a specified time period.

Antibiotic Exclusivity   Five years of additional market exclusivity granted by the FDA to qualifying antibiotics that treat serious or life-threatening infections, designed to incentivize development of new antimicrobial drugs to combat antibiotic resistance.

Approval   The regulatory process by which government agencies like the FDA review and authorize a drug for marketing and use by healthcare providers.

Asset   In biotech valuation context, a therapeutic candidate or product at any stage of development, from preclinical research through commercialization.

Average Sales Price (ASP)   The average price that manufacturers actually receive when they sell a drug, which Medicare uses to set reimbursement rates.

Average Wholesale Price (AWP)
An estimate of the average price retail pharmacies pay to wholesale distributors for prescription drugs.

Benchmark
A comparison standard or reference point used to evaluate performance, pricing, or other metrics against industry norms or competitors.

Best-in-Class
A drug that offers superior therapeutic benefits compared to existing treatments in the same category.

Biologics
Complex therapeutic products derived from living cells or organisms, including monoclonal antibodies, vaccines, recombinant proteins, RNA-based therapies, cell therapies, and gene therapies, which are typically manufactured through biotechnology processes and require specialized handling.

Biologics License Application (BLA)
The application submitted to the FDA seeking approval to market a biological product.

Biologics Price Competition and Innovation Act (BPCIA)
U.S. legislation establishing pathways for biosimilar approval while providing exclusivity protections for brand-name biologics.

Biomarkers
Measurable biological indicators used to assess disease progression, treatment response, or for patient selection in clinical trials.

Biosimilars
Highly similar versions of biologic medicines that demonstrate comparable efficacy, safety, and immunogenicity profiles to the reference product.

Biotech Company   A business that leverages living organisms, biological systems, and cellular processes to develop products and technologies that diagnose, treat, or prevent human diseases.

Breakthrough Therapy Designation   FDA designation for drugs that may offer substantial improvements over existing treatments, allowing for expedited development and review.

Capital Expenditure (CapEx)   Investments in long-term assets necessary for a company's growth and operations, such as buildings, equipment, or machinery.

CAR-T Therapy   Chimeric antigen receptor T-cell therapy that modifies a patient's T-cells to better fight cancer.

Cash Flows   The movement of money into and out of a business, representing actual receipts and payments rather than accounting profits.

Cell and Gene Therapy   A broad category of advanced biological treatments that encompasses both cell-based therapies (using living cells as medicine) and gene-based approaches (delivering genetic material), often combined in treatments like genetically modified CAR-T cells or gene-edited stem cells.

Clinical Development   The process of testing new drugs in human subjects through Phase 1, 2, and 3 trials to evaluate safety and efficacy.

Clinical Trial Phases   The sequential stages of human testing required for drug approval, progressing from small safety studies (Phase 1) to large-scale efficacy trials (Phase 3), followed by post-market monitoring (Phase 4).

Commercial Launch

The milestone event when a biotech product transitions from development to commercial availability in the marketplace following required regulatory approvals.

Commercialization

The systematic process of establishing a newly approved product in the marketplace, involving manufacturing scale-up, go-to-market strategy, distribution channel development, and achieving market penetration.

Contract Development and Manufacturing Organizations (CDMOs)

Third-party organizations that provide drug development and manufacturing services to biotech and pharma companies, from early-stage development through commercial production.

Cost of Goods Sold (COGS)

The direct costs attributable to producing a drug, including raw materials and manufacturing expenses.

Cumulative Probability of Success

The combined likelihood that a drug will advance through and complete all development stages up to reference point.

Depreciation and Amortization (D&A)

Accounting methods that spread the cost of long-term assets over time to reflect their gradual loss of value.

Diagnosis Rate

The percentage of patients with a particular disease who are correctly identified and diagnosed by healthcare providers.

Disclosure

The formal submission process where inventors record invention details and contributors to initiate the technology transfer process, essentially the first step researchers take to report their discoveries to the university.

Discounted Cash Flow (DCF)	A valuation method that estimates the value of an investment based on future cash flows adjusted for the time value of money and risks.
Discount Rate	The minimum return an investor expects for committing capital to a project, reflecting the cost of capital and investment risk.
Due Diligence	The investment evaluation process where potential investors thoroughly investigate a company's financial, scientific, and business aspects before making funding decisions.
Early-Mover	Pharmaceutical products that enter the market shortly after the first-in-class drug but ahead of later competitors.
Early-Stage Biotech	Companies focused primarily on research and development (preclinical or clinical stage) with no or limited commercial products, characterized by high risk and significant upside potential.
Effective Royalty Rate (EFR)	The weighted average royalty rate applied to a specific level of annual sales in tiered royalty structures.
Enterprise Value (EV)	The total value of the business, regardless of how it's financed (debt or equity).
Equity	Ownership stake in a company, representing the residual interest in assets after deducting liabilities, typically held by shareholders.
Fast Track Designation	FDA program designed to expedite development and review of drugs targeting serious conditions with high unmet medical needs.

Food and Drug Administration (FDA)	The U.S. federal agency responsible for regulating and approving drugs, biologics, and medical devices to ensure their safety and efficacy.
First-in-Class	The first drug approved in a new therapeutic category or with a novel mechanism of action.
First-Mover	The first pharmaceutical product to enter a particular therapeutic market, typically enjoying competitive advantages in a minimally differentiated market.
Generic or Biosimilar Competition	The market threat that occurs when patent protection or market exclusivity expires, allowing other companies to produce and sell equivalent versions of a drug at lower prices.
Generics	Therapeutically equivalent versions of brand-name small-molecule drugs that contain identical active ingredients and have proven bioequivalence and safety.
GLP-1 Receptor Agonists	A class of medications that mimic the hormone GLP-1 to regulate blood sugar levels and appetite.
Group Purchasing Organizations (GPOs)	Entities that aggregate purchasing power to negotiate lower drug prices and better contract terms on behalf of multiple healthcare entities, such as hospitals, health systems, and pharmacies.
Gross-to-Net	The reduction from a drug's list price (WAC) to the net revenue actually received by the manufacturer after accounting for rebates, discounts, chargebacks, and other pricing concessions.

Hatch-Waxman Act	U.S. legislation that established pathways for generic drug approval while providing patent protections for brand-name small-molecule drugs.
Immunotherapy	Treatments that harness or enhance the body's immune system to fight diseases, particularly cancer.
Incidence	The number of new cases of a disease that occur within a specific population during a defined time period.
Information and Communication Technologies (ICT)	The technology sector encompassing software, telecommunications, and digital services.
Intellectual Property (IP)	Legal rights protecting innovations, inventions, and creative works, including patents, trademarks, and trade secrets that are crucial for biotech companies.
Investigational New Drug (IND) Application	An application submitted to the FDA that must be approved before a new drug can be tested in humans during clinical trials.
Key Opinion Leaders (KOLs)	Influential physicians and researchers whose endorsement significantly impacts the adoption of new therapeutics.
Know-How	Tacit knowledge, methods, laboratory data, and confidential information that universities bundle with patents, including all the practical techniques and unwritten expertise that researchers developed to make their inventions work in real-world applications.
Label Expansion	The process of obtaining regulatory approval to market an existing drug for additional indications or patient populations.

Launch Curve   A model that predicts how a drug's sales will grow over time from market introduction to peak penetration.

Licensing   An agreement where one party grants rights to use intellectual property to another party in exchange for payments.

Loss of Exclusivity (LOE)   The point at which a brand-name drug's patent protection or market exclusivity expires, after which competitors can launch generic or biosimilar versions, typically resulting in significant price erosion and market share loss.

Machine Learning (ML)   Artificial intelligence-based platforms that accelerate drug discovery by leveraging algorithms to identify therapeutic candidates.

Market Exclusivity   The period during which a company has exclusive rights to market an approved drug before generic or biosimilar competition.

Market Share   The percentage of total sales or revenue that a company's product captures in a specific market over a set period.

Marketing   The strategic function focused on promoting pharmaceutical products to healthcare professionals and patients, encompassing market research, brand positioning, physician education, medical communications, and developing promotional strategies to drive product adoption and market share.

Mechanism of Action (MoA)   How a drug produces its intended effect in the body, critical for evaluating efficacy and safety profiles.

Medical Science Liaisons (MSLs)   Scientific professionals who bridge pharmaceutical companies and key opinion leaders, sharing clinical data.

Medication Possession Ratio (MPR)	A measure comparing the amount of medication dispensed to what a patient should have taken over a given period.
Mergers and Acquisitions (M&A)	Corporate transactions where companies combine or one company purchases another, common in biotech.
Milestone Payments	Payments made to licensors when specific predefined goals in drug development or commercialization are achieved.
Monoclonal Antibodies	Lab-made molecules designed to mimic the immune system's ability to fight harmful pathogens.
Monte Carlo Simulation	A computational technique that uses repeated random sampling from probability distributions to model uncertainty and generate a range of possible outcomes for complex financial or business scenarios.
Net Operating Profit After Tax (NOPAT)	A company's operating profit after deducting taxes, representing profit available to investors.
Net Present Value (NPV)	The total value of a project in today's dollars, calculated by discounting projected future cash flows to account for the time value of money.
New Chemical Entities (NCEs)	Products with active ingredients that have distinct chemical structures that have never before been approved by the FDA.
New Drug Application (NDA)	The application submitted to the FDA seeking approval to market a small-molecule drug.

New Molecular Entities (NMEs) Drugs containing active ingredients that have never been approved by the FDA for marketing in the U.S., representing novel therapeutic compounds rather than reformulations of existing drugs.

Orphan Drug A medication designated by the FDA to treat rare diseases affecting fewer than 200,000 people in the U.S., which receives 7 years of market exclusivity upon approval.

Patent Legal protection granting exclusive rights to an invention for a specified period, typically 20 years from filing date.

Patent Cost Reimbursement Licensee responsibility for ongoing patent prosecution and maintenance expenses, often including reimbursement of past university patent costs over several years.

Peak Market Penetration The maximum market share or sales volume a product achieves during its commercial lifecycle.

Pediatric Exclusivity An additional six months of market exclusivity granted by the FDA to companies that conduct required pediatric studies, extending existing patent or exclusivity periods as an incentive to test drugs in children.

Phase 1 Trial The first stage of human clinical testing, primarily focused on evaluating safety and determining appropriate dosing.

Phase 2 Trial The second stage of human clinical testing, evaluating a drug's effectiveness while continuing to monitor safety in larger patient groups.

Phase 3 Trial The third stage of human clinical testing, involving large-scale trials that compare experimental therapies with current standard treatments to confirm efficacy and monitor safety.

Phase 4 Trial Post-marketing surveillance studies conducted after approval to monitor long-term safety and effectiveness.

Pharmacy Benefit Managers (PBMs) Intermediary companies that negotiate drug prices with manufacturers, manage pharmacy networks, and administer prescription benefits for health plans and employers.

Pipeline A biotech company's portfolio of drug candidates currently in research and development, from preclinical studies through regulatory review but not yet commercially launched.

Portfolio A biotech company's complete collection of assets including marketed products, drugs in development, intellectual property, and other commercial assets across all therapeutic areas.

Post-money Valuation A company's value after receiving new investment, calculated as pre-money valuation plus the investment amount, used to determine investor ownership percentage.

Preclinical Development Laboratory and animal testing conducted before human trials to assess safety and biological activity.

Pre-money Valuation A company's value before receiving new investment, used as the baseline for calculating investor ownership stakes in funding rounds.

Prevalence The total number of people living with a disease at a given time, regardless of when diagnosed.

Primary Adherence (PA) The percentage of patients who fill their initial prescription after receiving it from a healthcare provider.

Priority Review   FDA designation shortening the review period to six months for drugs offering significant therapeutic advantages.

Probability of Success   The expected chance of successful progression through a clinical development phase based on industry benchmarks.

Research and Development (R&D)   The process of discovering, developing, and testing new drugs from initial research through clinical trials, representing the largest expense category for most biotech and pharma companies.

Retail Price (RP)   The average price charged to patients or end users for prescription medications at pharmacies.

Revenue   The total amount of money a company receives from selling its products or services before deducting any expenses.

Risk Adjustment   The process of modifying financial projections to account for the probability of success, particularly important in biotech valuation due to high development uncertainties.

Risk-adjusted NPV (rNPV) Method   A biotech valuation technique that applies probability of success adjustments at each development stage to the standard NPV calculation, accounting for clinical and regulatory risks to produce more realistic asset valuations.

RNA-based Therapies   Treatments that use RNA molecules such as mRNA, siRNA, or antisense oligonucleotides to modulate gene expression or protein production.

Royalties   Ongoing payments made to licensors based on a percentage of product sales, common in licensing agreements.

Sales
The revenue-generating function responsible for selling approved drugs or products to customers, including hospitals, pharmacies, distributors, and healthcare providers, typically involving specialized sales representatives who engage directly with prescribing physicians.

Seed Funding
The earliest stage of venture capital (VC) financing, used to fund initial product development and market research.

Selling, General, and Administrative (SG&A)
Operating expenses covering commercial operations (sales teams, marketing campaigns), corporate functions (executive leadership, finance, legal), and administrative overhead.

Series A Funding
The first significant round of venture capital (VC) financing, following seed funding, used to scale operations.

Serviceable Available Market (SAM)
The portion of the Total Addressable Market (TAM) that a biotech company can realistically capture with its specific product, considering factors like target patient population, competitive positioning, pricing strategy, and market access constraints.

Shape of the Curve
The steepness and trajectory of a drug's market uptake curve, describing how quickly and in what pattern the product gains market penetration from launch through peak adoption.

Small Business Innovation Research (SBIR) program
A federal grant program providing phased funding (Phase 1, 2, and 3) to qualify small businesses for innovative R&D projects that address federal agency needs and have commercial potential, with no partnership requirements.

Small Business Technology Transfer (STTR) program	A federal grant program requiring small businesses to formally partner with nonprofit research institutions (universities or federal laboratories) throughout the project lifecycle to develop innovative technologies for commercialization.
Small Molecules	Low molecular weight therapeutic compounds that are synthetically manufactured and designed to bind with specific biological targets (proteins, enzymes, receptors) to achieve therapeutic effects.
Stakeholders	Various interested parties in a biotech company, including investors, employees, patients, regulators, and partners who are affected by or can influence the company's decisions.
Standard of Care	The current accepted treatment approach for a particular condition, serving as the benchmark for new therapies.
Sum-of-the-Parts (SOTP) Method	A valuation approach that breaks down a biotech company into individual components, values each asset separately using appropriate methodologies, then combines these values with company-level factors to calculate the total company value.
Target Validation	The process of scientifically confirming that a specific biological target (protein, gene, or pathway) plays a causal role in disease progression and can be safely and effectively modulated by therapeutic intervention.
Technology Transfer Office (TTO)	University departments responsible for helping researchers turn their discoveries into commercial products by managing patents, finding companies to license the technology, and negotiating agreements.

Technology Transfer Process	The step-by-step pathway universities use to move inventions from initial disclosure by researchers through evaluation, patenting, marketing to companies, and final licensing agreements.
Therapeutic Area (TA)	A broad field of medicine focusing on treating related groups of diseases or medical conditions.
Therapeutic Indication (TI)	The specific use of a drug to treat a particular condition, requiring FDA approval for each indication.
Therapeutic Modality (TM)	The type of treatment approach or drug platform used to address disease, including small molecules, biologics (antibodies, proteins), cell therapies, gene therapies, and nucleic acid medicines (RNA, DNA).
Time to Peak	The duration required for a drug to achieve maximum market penetration or revenue after launch.
Total Addressable Market (TAM)	The total revenue opportunity available for a product or drug if it achieved 100% market penetration within its defined target market, representing the theoretical maximum market size.
Total Addressable Patient Population (TAPP)	The total number of patients a drug or product can realistically serve after accounting for disease segmentation, diagnosis rates, treatment rates, access rates (insurance, geography), and patient adherence/compliance rates.
Treatment Rate	The percentage of diagnosed patients within a target population who are eligible for and are recommended to receive medical treatment from healthcare providers.
Upfront Payment	Immediate compensation paid to licensors at the start of a licensing agreement.

Valuation   The process of determining the current worth or value of a company, asset, or investment based on various financial and strategic factors.

Venture Capital (VC)   Investment funding provided by professional investment firms to early-stage, high-growth potential companies in exchange for equity stakes, typically involving active involvement in strategic guidance and business development.

Weighted Average Cost of Capital (WACC)   The minimum rate of return a company must earn on its investments to satisfy all capital providers (debt and equity holders), serving as the standard discount rate for net present value calculations.

Wholesale Acquisition Cost (WAC)   The "list price" at which manufacturers sell drugs to wholesalers before any discounts or rebates.

Working Capital   The funds a company needs for daily operations, calculated as current assets minus current liabilities.

References

Chapter 1

Brex. 2025. "The Founder's Guide to Successful Startup Fundraising." *Brex*, January 16. https://www.brex.com/spend-trends/startup/startup-fundraising.

Brown, Dean G., Heike J. Wobst, Abhijeet Kapoor, Leslie A. Kenna, and Noel Southall. 2022. "Clinical Development Times for Innovative Drugs." *Nature Reviews Drug Discovery* 21 (11): 793–794. https://doi.org/10.1038/d41573-021-00190-9.

Chitale, Sadhana, Colm Lawler, and Arthur Klausner. 2022. "So You Want to Start a Biotech Company." *Nature Biotechnology* 40 (3): 296–300. https://doi.org/10.1038/s41587-022-01239-9.

DiMasi, Joseph A., Henry G. Grabowski, and Ronald W. Hansen. 2016. "Innovation in the Pharmaceutical Industry: New Estimates of R&D Costs." *Journal of Health Economics* 47: 20–33. https://doi.org/10.1016/j.jhealeco.2016.01.012.

Grabowski, Henry, Genia Long, Richard Mortimer, and Mehmet Bilginsoy. 2021. "Continuing Trends in US Brand-Name and Generic Drug Competition." *Journal of Medical Economics* 24 (1): 908–917. https://doi.org/10.1080/13696998.2021.1952795.

Jayasundara, Kavisha, Aidan Hollis, Murray Krahn, Muhammad Mamdani, Jeffrey S. Hoch, and Paul Grootendorst. 2019. "Estimating the Clinical Cost of Drug Development for Orphan Versus Non-Orphan Drugs." *Orphanet Journal of Rare Diseases* 14: 1–10. https://doi.org/10.1186/s13023-018-0990-4.

JPMorgan Chase & Co. 2023. *2023 Annual Biopharma Licensing and Venture Report.* JPMorgan Chase & Co, December. https://www.scribd.com/document/811498393/jpmorgan-dec-2023-biopharma-licensing-and-venture-report.

JPMorgan Chase & Co. 2025. *Biopharma Q1 2025: Strategic shifts and market dynamics.* JPMorgan Chase & Co, May. https://www.jpmorgan.com/content/dam/jpmorgan/documents/cb/insights/outlook/jpm-biopharma-deck-q1-2025-final-ada.pdf.

Murphey, Richard. 2023. "Top Biotech Venture Capital Funds of 2018–2023." *BayBridge Bio,* April 28. https://www.baybridgebio.com/blog/top_vcs_2018.html.

Paul, Steven M., Daniel S. Mytelka, Christopher T. Dunwiddie, Charles C. Persinger, Bernard H. Munos, Stacy R. Lindborg, and Aaron L. Schacht. 2010. "How to Improve R&D Productivity: The Pharmaceutical Industry's Grand Challenge." *Nature Reviews Drug Discovery* 9 (3): 203–214. https://doi.org/10.1038/nrd3078.

QLS Advisors. 2021. *Clinical Development Success Rates and Contributing Factors 2011–2020.* QLS Advisors, February 2021. https://go.bio.org/rs/490-EHZ-999/images/ClinicalDevelopmentSuccessRates2011_2020.pdf.

Robey, Seth, and Frank S. David. 2017. "Drug Launch Curves in the Modern Era." *Nature Reviews Drug Discovery* 16 (1): 13–14. https://doi.org/10.1038/nrd.2016.236.

Rome, Benjamin N., ChangWon C. Lee, and Aaron S. Kesselheim. 2021. "Market Exclusivity Length for Drugs with New Generic or Biosimilar Competition, 2012–2018." *Clinical Pharmacology & Therapeutics* 109 (2): 367–371. https://doi.org/10.1002/cpt.1983.

Schlander, Michael, Karla Hernandez-Villafuerte, Chih-Yuan Cheng, Jorge Mestre-Ferrandiz, and Michael Baumann. 2021. "How Much Does It Cost to Research and Develop a New Drug? A Systematic Review and Assessment." *Pharmacoeconomics* 39: 1243–1269. https://doi.org/10.1007/s40273-021-01065-y.

Sertkaya, Aylin, Trinidad Beleche, Amber Jessup, and Benjamin D. Sommers. 2024. "Costs of Drug Development and Research and Development Intensity in the US, 2000–2018." JAMA Network Open 7 (6): e2415445. https://doi.org/10.1001/jamanetworkopen.2024.15445.

Teramae, Fumio, Tomohiro Makino, Shintaro Sengoku, Yeongjoo Lim, Takashi Natori, and Kota Kodama. 2020. "Research on Pharmaceutical Product Life Cycle Patterns for Sustainable Growth." Sustainability 12 (21): Article 8938. https://doi.org/10.3390/su12218938.

University Lab Partners. 2021. "Funding Options for Biotech Startups." University Lab Partners Blog, June 9. https://www.universitylabpartners.org/blog/funding-options-for-biotech-startups.

U.S. Food and Drug Administration. 2014. "Types of Applications." U.S. Food and Drug Administration, October 23. https://www.fda.gov/drugs/how-drugs-are-developed-and-approved/types-applications.

U.S. Food and Drug Administration. 2020. "Frequently Asked Questions on Patents and Exclusivity." U.S. Food and Drug Administration, February 5. https://www.fda.gov/drugs/development-approval-process-drugs/frequently-asked-questions-patents-and-exclusivity.

U.S. Food and Drug Administration. 2023. "Fast Track, Breakthrough Therapy, Accelerated Approval, Priority Review." U.S. Food and Drug Administration, June 12. https://www.fda.gov/patients/learn-about-drug-and-device-approvals/fast-track-breakthrough-therapy-accelerated-approval-priority-review.

VCL Solutions. 2020. "Battling Maturity in the Pharma Product Life Cycle." VCL Solutions, December 4. https://vcl.solutions/insights/battling-maturity-in-the-pharma-product-life-cycle/.

Wang, Bo, Jun Liu, and Aaron S. Kesselheim. 2015. "Variations in Time of Market Exclusivity among Top-Selling Prescription Drugs in the United States." JAMA *Internal Medicine* 175 (4): 635–637. https://doi.org/10.1001/jamainternmed.2014.7968.

World Intellectual Property Organization. 2024. "Global Innovation Tracker." *World Intellectual Property Organization.* https://www.wipo.int/web-publications/global-innovation-index-2024/en/global-innovation-tracker.html.

Chapter 2

Aggarwal, Gautam, Barrett Rankin, and Andrew Lin. 2017. "Pharmaceutical Commercial Forecasting: Distilling Insights from Primary Research." Insights & Perspectives Series. Triangle Insights Group. https://triangleinsightsgroup.com/wp-content/uploads/2018/01/Pharmaceutical_Commercial_Forecasting-2.pdf.

American Cancer Society. 2022. Breast Cancer Facts & Figures 2022–2024. American Cancer Society. https://www.cancer.org/content/dam/cancer-org/research/cancer-facts-and-statistics/breast-cancer-facts-and-figures/2022-2024-breast-cancer-fact-figures-acs.pdf.

American Diabetes Association. 2023. "Standards of Care in Diabetes—2023 Abridged for Primary Care Providers." *Clinical Diabetes* 41(1): 4–31. https://doi.org/10.2337/cd23-as01.

Arnold, Michael J. 2021. "Venous Thromboembolism: Management Guidelines from the American Society of Hematology." *American Family Physician* 104(4): 429–431. https://www.aafp.org/pubs/afp/issues/2021/1000/p429.html.

Birt, Julie, Joseph Johnston, and David Nelson. 2014. "Exploration of Claims-Based Utilization Measures for Detecting Potential Nonmedical Use of Prescription Drugs." *Journal of Managed Care Pharmacy* 20: 639–646. https://doi.org/10.18553/jmcp.2014.20.6.639.

Blaschke, Terrence F., Lars Osterberg, Bernard Vrijens, and John Urquhart. 2012. "Adherence to Medications: Insights Arising from Studies on the Unreliable Link Between Prescribed and Actual Drug Dosing Histories." *Annual Review of Pharmacology and Toxicology* 52(1): 275–301. https://doi. org/10.1146/annurev-pharmtox-011711-113247.

Bruen, Brian K., and Katherine Young. 2014. "Paying for Prescribed Drugs in Medicaid: Current Policy and Upcoming Changes." *Kaiser Family Foundation*, May 23. https://www.kff.org/medicaid/issue-brief/paying-for-prescribed-drugs-in-medicaid-current-policy-and-upcoming-changes/.

California Department of Health Care Access and Information. 2025. "Cost Transparency." *California Department of Health Care Access and Information.* https://www.hcai.ca.gov/data/cost-transparency/.

Centers for Disease Control and Prevention. 2024. *National Diabetes Statistics Report.* Centers for Disease Control and Prevention, May 15. https://www.cdc.gov/diabetes/php/data-research/index.html.

Centers for Medicare & Medicaid Services. 2025. "ASP Pricing Files" CMS. *gov*, June. https://www.cms.gov/medicare/payment/part-b-drugs/asp-pricing-files.

Cha, Myoung, and Flora Yu. 2014. "Pharma's First-to-Market Advantage." *McKinsey & Company*, September 1. https://www.mckinsey.com/industries/life-sciences/our-insights/pharmas-first-to-market-advantage.

Cho, Sung Joon, Jungmee Kim, Jin Yong Lee, and Jee Hoon Sohn. 2022. "Adherence to Antipsychotic Drugs by Medication Possession Ratio for Schizophrenia and Similar Psychotic Disorders in the Republic of Korea: A Retrospective Cohort Study." *Clinical Psychopharmacology and Neuroscience* 20(3): 491. https://doi.org/10.9758/cpn.2022.20.3.491.

Curtiss, Frederic R. 2010. "What Is the Price Benchmark to Replace Average Wholesale Price (AWP)?" *Journal of Managed Care Pharmacy* 16(7): 492–501. https://doi.org/10.18553/jmcp.2010.16.7.492.

Elkout, Hajer, Peter J. Helms, Colin R. Simpson, and James S. McLay. 2012. "Adequate Levels of Adherence with Controller Medication Is Associated with Increased Use of Rescue Medication in Asthmatic Children." PLoS One 7(6): e39130. https://doi.org/10.1371/journal.pone.0039130.

Ellegård, Sander, Kristina Engvall, Mustafa Asowed, Anna-Lotta Hallbeck, Nils Elander, and Olle Stål. 2022. "Long-Term Follow-Up of Early Stage HER2-Positive Breast Cancer Patients Treated with Trastuzumab: A Population-Based Real World Multicenter Cohort Study." *Frontiers in Oncology* 12: 861324. https://doi.org/10.3389/fonc.2022.861324.

Engmann, Natalie J., Danny Sheinson, Komal Bawa, Carmen Ng, and Gabriel Pardo. 2021. "Persistence and Adherence to Ocrelizumab Compared with Other Disease-Modifying Therapies for Multiple Sclerosis in US Commercial Claims Data." *Journal of Managed Care & Specialty Pharmacy* 27(5): 639–649. https://doi.org/10.18553/jmcp.2021.20413.

Evans, Charity, Ruth Ann Marrie, Shenzhen Yao, Karen M. Yeung, Charity Evans, Helen Tremlett, James M. Bolton, et al. 2021. "Medication Adherence in Multiple Sclerosis as a Potential Model for Other Chronic Diseases: A Population-Based Cohort Study." BMJ *Open* 11 (2): e043930. https://doi.org/10.1136/bmjopen-2020-043930.

Fein, Adam. 2012. "Surprise? CMS Computes and Publishes Pharmacy Prescription Profit Margins." *Drug Channels*, December 11. https://www.drugchannels.net/2012/12/surprise-cms-computes-and-publishes.html.

Fein, Adam. 2022. "Warped Incentives Update: The Gross-to-Net Bubble Exceeded $200 Billion in 2021." *Drug Channels*, March 22. https://www.drugchannels.net/2022/03/warped-incentives-update-gross-to-net.html.

Fein, Adam. 2024a. "Gross-to-Net Bubble Update: 2023 Pricing Realities at 10 Top Drugmakers." *Drug Channels*, July 23. https://www.drugchannels.net/2024/07/gross-to-net-bubble-update-2023-pricing.html.

Fein, Adam. 2024b. "PBM Power: The Gross-to-Net Bubble Reached $334 Billion in 2023—but Will Soon Start Deflating." *Drug Channels*, July 16. https://www.drugchannels.net/2024/07/pbm-power-gross-to-net-bubble-reached.html.

Financial Times. 2023. "The Battle Over European Drug Pricing." *Financial Times*, January 19. https://www.ft.com/content/fe8e3e11-5f08-46be-8061-7d76674e0a60.

First Databank. 2025. "Drug Databases That Empower Critical Decisions." *First Databank*. https://www.fdbhealth.com.

FitzGerald, J. D., Nicola Dalbeth, Ted Mikuls, Tuhina Neogi, Robert Terkeltaub, John K. O'Dell, Kenneth G. Saag, et al. 2020. "2020 American College of Rheumatology Guideline for the Management of Gout." *Arthritis Care & Research* 72 (6): 744–760. https://doi.org/10.1002/acr.24180.

Fleming, Thomas. 2005. *Red Book*. Montvale, NJ: Thomson Healthcare.

Fram, Francine, Svetlana Gogolina. 2023. "A Guide to Better Demand Research." *Ipsos*, May. https://www.ipsos.com/sites/default/files/ct/publication/documents/2023-05/Ipsos%20PoV_A%20Guide%20to%20Better%20Demand%20Research%20-%20May%202023.pdf.

Gemmete, J.J., and S.K. Mukherji. 2011. "Trastuzumab (Herceptin)." *American Journal of Neuroradiology* 32(8): 1373–1374. https://doi.org/10.3174/ajnr.a2619.

Genentech. 2023. "Herceptin (Trastuzumab) Dosing in HER2+ Adjuvant Breast Cancer." *Genentech*, July 19. https://www.herceptin.com/hcp/adjuvant-breast-cancer/dosing-and-administration/dosing.html.

George, Peter M., Paolo Spagnolo, Michael Kreuter, Vincent Cottin, Toby M. Maher, Marlies S. Wijsenbeek, Kevin K. Brown, et al. 2020. "Progressive Fibrosing Interstitial Lung Disease: Clinical Uncertainties, Consensus Recommendations, and Research Priorities." *The Lancet Respiratory Medicine* 8 (9): 925–934. https://doi.org/10.1016/S2213-2600(20)30355-6.

Guy, Ann. 2022. "Generic Drugs: Are They on a Par with Pricier Brands?" UCSF *Magazine*, Winter. https://magazine.ucsf.edu/generic-drugs-are-they-par-pricier-brands.

Haque, Reina, Jiaxiao Shi, Joanie Chung, Debra P. Ritzwoller, Erin J. Aiello Bowles, Leslie A. Walker, Ted H. Kim, et al. 2017. "Medication Adherence, Molecular Monitoring, and Clinical Outcomes in Patients with Chronic Myelogenous Leukemia in a Large HMO." *Journal of the American Pharmacists Association* 57 (3): 303–310. https://doi.org/10.1016/j.japh.2017.02.013.

Health Action International. 2019. "Price & Availability Data." July 29. *Health Action International*. https://www.haiweb.org/what-we-do/price-availability-affordability/price-availability-data/.

Higuera, Lucas, Caroline S. Carlin, and Sarah Anderson. 2016. "Adherence to Disease-Modifying Therapies for Multiple Sclerosis." *Journal of Managed Care & Specialty Pharmacy* 22(12): 1394–1401. https://doi.org/10.18553/jmcp.2016.22.12.1394.

Hoffmann, Olaf, Friedemann Paul, Rocco Haase, Raimar Kern, and Tjalf Ziemssen. 2024. "Preferences, Adherence, and Satisfaction: Three Years of Treatment Experiences of People with Multiple Sclerosis." *Patient Preference and Adherence*: 455–466. https://doi.org/10.2147/ppa.s452849.

Hungin, A. Pali S., Catherine Hill, Michael Molloy-Bland, and Anan Raghunath. 2012. "Systematic Review: Patterns of Proton Pump Inhibitor Use and Adherence in Gastroesophageal Reflux Disease." *Clinical Gastroenterology and Hepatology* 10(2): 109–116. https://doi.org/10.1016/j.cgh.2011.07.008.

Iglay, Kristy, Xiting Cao, Panagiotis Mavros, Kruti Joshi, Shengsheng Yu, and Kaan Tunceli. 2015. "Systematic Literature Review and Meta-Analysis of Medication Adherence with Once-Weekly Versus Once-Daily Therapy." *Clinical Therapeutics* 37(8): 1813–1821. https://doi.org/10.1016/j.clinthera.2015.05.505.

IQVIA. 2023. *The Use of Medicines in the U.S. 2023. Usage and Spending Trends and Outlook to 2027.* IQVIA, April. https://www.iqvia.com/insights/the-iqvia-institute/reports-and-publications/reports/the-use-of-medicines-in-the-us-2023.

IQVIA. 2024. *The Use of Medicines in the U.S. 2024: Usage and Spending Trends and Outlook to 2028.* IQVIA, April. https://www.iqvia.com/insights/the-iqvia-institute/reports-and-publications/reports/the-use-of-medicines-in-the-us-2024.

Jacquet, Pierre, TJ Bilodeau, and Adam Nover. 2024. "First vs. Best in Class – Simplifying the Equation for Biopharma." *L.E.K. Consulting*, June 27. https://www.lek.com/insights/hea/us/ei/first-vs-best-class-simplifying-equation-biopharma.

Jensen, Frodi Fridason, Kjell E J Håkansson, Britt Overgaard Nielsen, Ulla Møller Weinreich, and Charlotte Suppli Ulrik. 2021. "Self-Reported vs. Objectively Assessed Adherence to Inhaled Corticosteroids in Asthma." *Asthma Research and Practice* 7(1): 7. https://doi.org/10.1186/s40733-021-00072-2.

Kim, Yundeok, Tae-Hwa Go, Jaeyeon Jang, Soo-Jeong Kim, Jae-Cheol Jo, Sung-Hyun Kim, Ji-Young Park, et al. 2021. "Survival Impact of Adherence to Tyrosine Kinase Inhibitor in Chronic Myeloid Leukemia." *The Korean Journal of Internal Medicine* 36 (6): 1450–1460. https://doi.org/10.3904/kjim.2020.537.

Leapman, Michael S., and Peter R. Carroll. 2017. "What Is the Best Way Not to Treat Prostate Cancer?" *Urologic Oncology: Seminars and Original Investigations* 35(2). https://doi.org/10.1016/j.urolonc.2016.09.003.

Liss, David T., Manisha Cherupally, Matthew J. O'Brien, Olivia S. Lee, Stephanie M. Zuckerman, Kenzie A. Cameron, Jason N. Doctor, et al. 2023. "Treatment Modification After Initiating Second-Line Medication for Type 2 Diabetes." *American Journal of Managed Care* 29 (12). https://doi.org/10.37765/ajmc.2023.89466.

Maine Health Data Organization. 2025. *Prescription Drug Pricing Transparency Report*. Maine Health Data Organization. https://mhdo.maine.gov/RxDrugPricingTransparency.htm.

Maniero, Carmela, Aleksandra Lopuszko, Kyriaki-Barbara Papalois, Ajay Gupta, Vikas Kapil, and Mohammed Y Khanji. 2023. "Non-Pharmacological Factors for Hypertension Management: A Systematic Review of International Guidelines." *European Journal of Preventive Cardiology* 30(1): 17–33. https://doi.org/10.1093/eurjpc/zwac163.

Mauboussin, Michael, and Dan Callahan. 2015. *Total Addressable Market: Methods to Estimate a Company's Potential Sales*. Credit Suisse, September. https://strawman.com/member/uploads/objects/55/ff4dd2946e4c06eb09aec2c537921b21f7afff.pdf.

McNaughton, Candace D., Peter C. Austin, Cynthia A. Jackevicius, Dennis T. Ko, Tara Gomes, Paul J. Youngson, Robert A. Fowler, et al. 2025. "Incident Prescriptions for Common Cardiovascular Medications: Comparison of Recent Versus Pre-2020 Medication Adherence and Discontinuation in Three Universal Health Care Systems." *BMC Cardiovascular Disorders* 25: 82. https://doi.org/10.1186/s12872-025-04492-3.

Medicaid. 2025. "Pharmacy Pricing." *Medicaid.gov*, June. https://www.medicaid.gov/medicaid/prescription-drugs/pharmacy-pricing/index.html.

MedPAC. 2019. "Improving Medicare's Payment for Part B Drugs: Requiring Pharmaceutical Manufacturer Reporting of Sales Price Data." *MedPAC*, June 14. https://www.medpac.gov/improving-medicares-payment-for-part-b-drugs-requiring-pharmaceutical-manufacturer-reporting-of-sales-price-data/#:~:text=,ASP%20data%20and%20set%20payments.

Micromedex Solutions. 2025. "Micromedex." *Merative*, June. https://www.micromedexsolutions.com.

Mulcahy, Andrew W., Daniel Schwam, and Susan L. Lovejoy. 2024. *International Prescription Drug Price Comparisons: Estimates Using 2022 Data*. RAND, February 1. https://doi.org/10.7249/rra788-3.

Park, Yoonyoung, Hyuna Yang, Amar K Das, and Gigi Yuen-Reed. 2018. "Prescription Fill Rates for Acute and Chronic Medications in Claims-EMR Linked Data." *Medicine* 97(44): e13110. https://doi.org/10.1097/md.0000000000013110.

Perehudoff, Katrina, Kaitlin Mara, and Ellen 't Hoen. 2021. *What Is the Evidence on Legal Measures to Improve the Transparency of Markets for Medicines, Vaccines and Other Health Products (World Health Assembly Resolution WHA72.8)?* World Health Organization. https://www.ncbi.nlm.nih.gov/books/NBK572572/.

Pereira, Stephen P., Lucy Oldfield, Alexander Ney, Diane Hart, William Keane, Hemant Kocher, Mark Lawler, et al. 2020. "Early Detection of Pancreatic Cancer." *The Lancet Gastroenterology & Hepatology* 5 (7): 698–710. https://doi.org/10.1016/S2468-1253(20)30072-5.

Piragine, Eugenia, Davide Petri, Alma Martelli, Vincenzo Calderone, and Ersilia Lucenteforte. 2023. "Adherence to Oral Antidiabetic Drugs in Patients with Type 2 Diabetes: Systematic Review and Meta-Analysis." *Journal of Clinical Medicine* 12(5): 1981. https://doi.org/10.3390/jcm12051981.

Porath, Daniel. 2018. "Size and Dynamics of Order-of-Entry Effects in Pharmaceutical Markets." *International Journal of Market Research* 60(1): 50–66. https://doi.org/10.1177/1470785317744669.

Prescription Analytics. 2025. "Key Terms in Pharmaceutical Government Pricing." *Prescription Analytics Inc.* https://prescriptionanalytics.com/white-paper/key-terms-in-pharmaceutical-government-pricing/.

Roche. 2025. "Herceptin Dosing and Administration Recommendations." *Roche.* https://www.medically.roche.com/global/en/medinfo/kadcyla/herceptin-dosing-and-administration-recommendations.html.

SEER. 2025. "Cancer of the Breast (Female) – Cancer Stat Facts." *National Cancer Institute, Surveillance, Epidemiology, and End Results Program (SEER).* https://www.seer.cancer.gov/statfacts/html/breast.html.

Schulze, Ulrik, and Michael Ringel. 2013. "What Matters Most in Commercial Success: First-in-Class or Best-in-Class?" *Nature Reviews Drug Discovery* 12(6): 419. https://doi.org/10.1038/nrd4035.

Schumacher, Leslie, and Steven Miller. 2021. "Red Book: Fact or Fiction." *MSP Network*. September 29. https://cdn.ymaws.com/mspnetwork.org/resource/collection/A0D530B4-59AC-4E26-AAB5-A9B426C64AC0/Red_Book_Fact_or_Fiction.pdf.

SSR Health. 2025. "US Brand Rx Net Pricing Tool." SSR *Health*. https://www.ssrhealth.com/dataset/.

Tamblyn, Robyn, Tewodros Eguale, Allen Huang, Nancy Winslade, and Pamela Doran. 2014. "The Incidence and Determinants of Primary Nonadherence with Prescribed Medications in Primary Care: A Cohort Study." *Annals of Internal Medicine* 160: 441–450. https://doi.org/10.7326/m13-1705.

Tang, Karen L., Hude Quan, and Doreen M. Rabi. 2017. "Measuring Medication Adherence in Patients with Incident Hypertension: A Retrospective Cohort Study." BMC *Health Services Research* 17: 1–16. https://doi.org/10.1186/s12913-017-2073-y.

Taylor, Phil. 2022. "Nice Backs Biogen's Vumerity for Multiple Sclerosis." *Pharmaphorum*, April 13. https://www.pharmaphorum.com/news/nice-backs-biogens-vumerity-for-multiple-sclerosis.

Trigger, Sarah, Xin Xu, Ann Malarcher, Esther Salazar, Hyungsik Shin, and Stephen Babb. 2023. "Trends in Over-the-Counter Nicotine Replacement Therapy Sales, US, 2017–2020." *American Journal of Preventive Medicine* 64(5): 650–657. https://doi.org/10.1016/j.amepre.2022.12.008.

Trinity Life Sciences. 2023. "Japan's Latest Drug Pricing Policy Updates: Key Changes in 2022 and the Expected Impact." *Trinity Life Sciences*, September 28. https://www.trinitylifesciences.com/blog/japans-latest-drug-pricing-policy-updates-key-changes-and-expected-impact/.

U.K. Department of Health and Social Care. 2018. "Pharmaceutical Price Regulation Scheme 2014." GOV.UK, January 22. https://www.gov.uk/government/publications/pharmaceutical-price-regulation-scheme-2014.

U.S. Census Bureau. 2024. "New 2024 Population Estimates Show Nation's Population Grew by About 1% to 340.1 Million Since 2023." U.S. Census Bureau. December 19. https://www.census.gov/library/stories/2024/12/population-estimates.html.

U.S. Department of Health and Human Services. 2005. Medicaid Drug Price Comparison: Average Sales Price to Average Wholesale Price. U.S. Department of Health and Human Services, June 29. https://oig.hhs.gov/documents/evaluation/2311/OEI-03-05-00200-Complete%20Report.pdf.

U.S. Food and Drug Administration. 2023. "Improving Medication Adherence and Patient Experience by Researching Patient Perceptions of Generic Drugs." U.S. Food and Drug Administration, August 1. https://www.fda.gov/drugs/cder-conversations/improving-medication-adherence-and-patient-experience-researching-patient-perceptions-generic-drugs.

U.S. Inflation Calculator. 2025. "US Inflation Rates: 2000–2025." U.S. Inflation Calculator. https://www.usinflationcalculator.com/inflation/current-inflation-rates/.

U.S. Securities and Exchange Commission. 2025. "Edgar Advanced Search." U.S. Securities and Exchange Commission. https://www.sec.gov/edgar/search/.

Visram, Alissa, Joselle Cook, and Rahma Warsame. 2021. "Smoldering Multiple Myeloma: Evolving Diagnostic Criteria and Treatment Strategies." Hematology 2021(1): 673–681. https://doi.org/10.1182/hematology.2021000254.

Wang, Tao, Qian Liu, William Tjhioe, Lei Zhao, Yongjun Chen, Lijuan Zhao, Zhiping Wu, et al. 2017. "Therapeutic Potential and Outlook of Alternative Medicine for Osteoporosis." Current Drug Targets 18 (9): 1051–1068. https://doi.org/10.2174/1389450118666170123145713.

Weisser, Burkhard, Hans-Georg Predel, Anton Gillessen, Roland Schmieder, Dagmar Groth, Antje Bergmann, Bernd Griese, et al. 2020. "Single Pill Regimen Leads to Better Adherence and Clinical Outcome in Daily Practice in Patients Suffering from Hypertension and/or Dyslipidemia: Results of a Meta-Analysis." *High Blood Pressure & Cardiovascular Prevention* 27: 157–164. https://doi.org/10.1007/s40292-020-00363-3.

White, Lilian. 2022. "Osteoporosis Prevention, Screening, and Diagnosis: ACOG Recommendations." *American Family Physician* 106(5): 587–588. https://www.aafp.org/pubs/afp/issues/2022/1100/practice-guidelines-osteoporosis.html.

Wolters Kluwer. 2025. "Medi-Span Price Rx: Online Drug Pricing Tool." *Wolters Kluwer.* https://www.wolterskluwer.com/en/solutions/medi-span/price-rx.

Chapter 3

Aitken, M., M. Kleinrock, and E. Muñoz. 2020. Biosimilars in the United States 2020–2024: Competition, Savings, and Sustainability. IQVIA Institute for Human Data Science, October 29. https://www.bigmoleculewatch.com/wp-content/uploads/sites/2/2020/10/iqvia-institute-biosimilars-in-the-united-states.pdf.

Allmendinger, Andrea. 2021. "Opportunities in an Evolving Pharmaceutical Development Landscape: Product Differentiation of Biopharmaceutical Drug Products." Pharmaceutical Research 38 (5): 739–757. https://doi.org/10.1007/s11095-021-03037-5.

Association for Accessible Medicines. 2024. The U.S. Generic & Biosimilar Medicines Savings Report. Association for Accessible Medicines, September. https://accessiblemeds.org/wp-content/uploads/2025/01/AAM-2024-Generic-Biosimilar-Medicines-Savings-Report.pdf.

Bauer, Hans H., and Marc Fischer. 2000. "Product Life Cycle Patterns for Pharmaceuticals and Their Impact on R&D Profitability of Late Mover Products." International Business Review 9 (6): 703–725. https://doi.org/10.1016/s0969-5931(00)00028-7.

Biosimilars Forum. 2021. Biosimilars Have Significantly Lowered Prices of All Biologics. Biosimilars Forum. https://biosimforum.wpengine.com/wp-content/uploads/Xcenda-ASP-One-Pager.pdf.

BioSpace. 2021. "The Essential Guide to Drug Commercialization: A Manufacturer's Handbook for Selecting Solutions and Partners." BioSpace, July 22. https://www.biospace.com/the-essential-guide-to-drug-commercialization-a-manufacturer-s-handbook-for-selecting-solutions-and-partners.

Biotech Primer. 2023. "Drug Commercialization Process: From Research to Market." Biotech Primer, August 31. https://biotechprimer.com/drug-commercialization-process/.

Bloomberg Law. 2011. "The Biologics Price Competition and Innovation Act: The Pros and Cons of Biosimilar Approval." Bloomberg Law, February 7. https://news.bloomberglaw.com/pharma-and-life-sciences/the-biologics-price-competition-and-innovation-act-the-pros-and-cons-of-biosimilar-approval.

Callens, Sarah, Jahid Hasan, Ryan McCoy, and Stephen Ward. 2016. "Cell and Gene Therapy Manufacturing: The Necessity for a Cost-Based Development Approach." Cell Gene Therapy Insights 2(1): 115-120. https://www.insights.bio/immuno-oncology-insights/journal/article/521/cell-and-gene-therapy-manufacturing-the-necessity-for-a-cost-based-development-approach.

Celen, Wim, Gian-Carlo Walker, Surbhi Jain, and Arya Taupin. 2023. "A Strategic Approach to Cost in Biopharma." Boston Consulting Group, November 22. https://www.bcg.com/publications/2023/biopharma-manufacturing-cost-reduction.

Censora. 2025. "The Essential Guide to Drug Commercialization." Censora. https://www.cencora.com/resources/pharma/guide-to-drug-commercialization.

Cha, Myoung, and Flora Yu. 2014. "Pharma's First-to-Market Advantage." *McKinsey & Company*, September 1. https://www.mckinsey.com/industries/life-sciences/our-insights/pharmas-first-to-market-advantage.

Conrad, Ryan, Sarah Nance, Zachary Tillman, Kristin Davis. 2024. "Estimating Cost Savings from New Generic Drug Approvals in 2022." U.S. Food and Drug Administration, September. https://www.fda.gov/media/182435/download.

Donoghoe, Nicholas, Jon Duane, Jin Kim, and George Xu. 2017. "Pulling Away from the Pack in Drug Launches." Nature Reviews Drug Discovery 16 (11): 749–750. https://doi.org/10.1038/nrd.2017.122.

Dunleavy, Kevin. 2024. "Sanofi Completes Build Out of $554m Pandemic-Ready Modular Plant for Vaccines, Biologics." Fierce Pharma, September 12, 2024. https://www.fiercepharma.com/manufacturing/sanofi-completes-construction-554m-modular-plant-france-can-shift-production-during.

Frank, Richard G., Thomas G. McGuire, and Ian Nason. 2021. "The Evolution of Supply and Demand in Markets for Generic Drugs." The Milbank Quarterly 99 (3): 828–852. https://doi.org/10.1111/1468-0009.12517.

Goldman, Dana, and Tomas Philipson. 2021. "Biosimilars Competition Helps Patients More than Generic Competition." USC Schaeffer, October 8. https://healthpolicy.usc.edu/article/biosimilars-competition-helps-patients-more-than-generic-competition/.

Gooch, John, Frank Cordes, Frank Bressau, and Philip Berk. 2017. "What Does—and Does Not—Drive Biopharma Cost Performance." Boston Consulting Group, July 7. https://www.bcg.com/publications/2017/biopharmaceuticals-operations-what-does-and-does-not-drive-biopharma-cost-performance.

Goodarzi, Maryam, Mehdi Goodarzi, Mahdieh Sheikhi, and Mohammadmehdi Shabani. 2017. "Commercialization Process of Biopharmaceuticals Development." International Journal of Innovation, Management and Technology 8 (4): 289–292. https://doi.org/10.18178/ijimt.2017.8.4.743.

Grabowski, Henry, Genia Long, Richard Mortimer, and Mehmet Bilginsoy. 2021. "Continuing Trends in US Brand-Name and Generic Drug Competition." Journal of Medical Economics 24 (1): 908–917. https://doi.org/10.1080/13696 998.2021.1952795.

IQVIA. 2024. The Global Use of Medicines 2024: Outlook to 2028. IQVIA, January. https://www.iqvia.com/insights/the-iqvia-institute/reports-and-publications/reports/the-global-use-of-medicines-2024-outlook-to-2028.

Jiang, John, Jing Kong, and Joseph Grogan. 2021. "How Did the Public U.S. Drugmakers' Sales, Expenses and Profits Change over Time?" USC Schaeffer, November 5. https://healthpolicy.usc.edu/evidence-base/how-did-the-public-u-s-drugmakers-sales-expenses-and-profits-change-over-time/.

Ledley, Fred D., Sarah Shonka McCoy, Gregory Vaughan, and Ekaterina Galkina Cleary. 2020. "Profitability of Large Pharmaceutical Companies Compared with Other Large Public Companies." JAMA 323 (9): 834–843. https://doi.org/10.1001/jama.2020.0442.

Macdonald, Gareth John. 2024. "Cell & Gene Therapy Costs Drive Deals." GEN - Genetic Engineering and Biotechnology News, May 1. https://www.genengnews.com/topics/bioprocessing/cell-gene-therapy-costs-drive-deals/.

Mandell, Emily, and Joshua Hattem. 2019. "The Growing Challenge of Product Differentiation." ZS, April 16. https://www.zs.com/insights/the-growing-challenge-of-product-differentiation.

Merrill, Jessica. 2022. "The Next Big Patent Cliff Is Coming, and Time Is Running Out to Pad the Fall." Citeline, April 4. https://scrip.citeline.com/SC146175/The-Next-Big-Patent-Cliff-Is-Coming-And-Time-Is-Running-Out-To-Pad-The-Fall.

Mittal, Bhavishya. 2024. "Evaluation of Pharmaceutical Drug Development Lifecycle." In Sustainable Global Health Systems and Pharmaceutical Development, 125–168. Cham: Springer International Publishing. https://doi.org/10.1007/978-3-031-50415-0_5.

Morrison, Anton. 2020. "The Evolving World of Pharma Marketing." Pharmacy Times, May 28. https://www.pharmacytimes.com/view/the-evolving-world-of-pharma-marketing.

Patel, Rinku. 2023. "Exclusivity–Which One Is for Me?" U.S. Food and Drug Administration. https://www.fda.gov/media/135234/download.

Pierre, Colette. 2022. "2021 Pharma Statistics: Long-Term Cost and Underlying EBIT Analysis." Hardman & Co., June 6. https://hardmanandco.com/2021-pharma-statistics-long-term-cost-underlying-ebit-analysis/.

Positano, Lorenzo, Lu Chen, Yane Yane Cheng, and Priyanka Aggarwal. 2019. "Getting a Grip on COGS in Generic Drugs." Boston Consulting Group, July 1. https://www.bcg.com/publications/2019/getting-a-grip-on-cogs-in-generic-drugs.

Pearson, Steven, Maria Low, Adrian Towse, Celia S. Segel, and Chris Henshall. 2020. Cornerstones of 'Fair' Drug Coverage: Appropriate Cost Sharing and Utilization Management Policies for Pharmaceuticals. Institute for Clinical and Economic Review, September 28. https://icer.org/wp-content/uploads/2020/11/Cornerstones-of-Fair-Drug-Coverage-_-September-28-2020-corrections-1-5-21.pdf.

PharmExec. 2020. "Managing Product Lifecycle." PharmExec, November 15. https://www.pharmexec.com/view/managing-product-lifecycle.

PhRMA. 2024. "40 Years of Hatch-Waxman: What Is the Hatch-Waxman Act?" PhRMA Blog, September 19. https://phrma.org/Blog/40-Years-of-Hatch-Waxman-What-is-the-Hatch-Waxman-Act.

Porath, Daniel. 2018. "Size and Dynamics of Order-of-Entry Effects in Pharmaceutical Markets." International Journal of Market Research 60 (1): 50–66. https://doi.org/10.1177/1470785317744669.

Robey, Seth, and Frank S. David. 2017. "Drug Launch Curves in the Modern Era." Nature Reviews Drug Discovery 16 (1): 13–14. https://doi.org/10.1038/nrd.2016.236.

Rome, Benjamin N., ChangWon C. Lee, and Aaron S. Kesselheim. 2021. "Market Exclusivity Length for Drugs with New Generic or Biosimilar Competition, 2012–2018." Clinical Pharmacology & Therapeutics 109 (2): 367–371. https://doi.org/10.1002/cpt.1983.

Rudge, Dean. 2022. "Humira One Year Out: The Largest LOE Event in US Pharma History." Generics Bulletin, January 31. https://generics.citeline.com/GB151601/Humira-One-Year-Out-The-Largest-LOE-Event-In-US-Pharma-History.

Saikia, Dibyajyoti, Hansraj Kumar, and Subodh Kumar. 2023. "Impact of COVID-19 on Drug Discovery and Development: A Pharmacologist's Perspective." Advanced Biomedical Research 12 (1): 212. https://doi.org/10.4103/abr.abr_174_23.

Samsung Bioepis. 2025. Biosimilar Market Report. 9th edition, Q2 2025. Samsung Bioepis Co., March. https://www.samsungbioepis.com/upload/attach/SB+Biosimilar+Market+Report+Q2+2025.pdf.

Teramae, Fumio, Tomohiro Makino, Shintaro Sengoku, Yeongjoo Lim, Takashi Natori, and Kota Kodama. 2020. "Research on Pharmaceutical Product Life Cycle Patterns for Sustainable Growth." Sustainability 12 (21): Article 8938. https://doi.org/10.3390/su12218938.

U.S. Food and Drug Administration. 2020. "Frequently Asked Questions on Patents and Exclusivity." U.S. Food and Drug Administration, February 5. https://www.fda.gov/drugs/development-approval-process-drugs/frequently-asked-questions-patents-and-exclusivity#howlongpatentterm.

U.S. Food and Drug Administration. 2022. "The Generic Drug Approval Process." U.S. *Food and Drug Administration*, March 17. https://www.fda.gov/drugs/cder-conversations/generic-drug-approval-process.

U.S. Food and Drug Administration. 2025. "Approved Drug Products with Therapeutic Equivalence Evaluations (Orange Book)." U.S. Food and Drug Administration, May 9. https://www.fda.gov/drugs/drug-approvals-and-databases/approved-drug-products-therapeutic-equivalence-evaluations-orange-book#Publications.

U.S. Patent and Trademark Office. 2025. "Online Patent Tools." U.S. Patent and Trademark Office. https://www.uspto.gov/patents/basics/online-patent-tools.

Wang, Bo, Jun Liu, and Aaron S. Kesselheim. 2015. "Variations in Time of Market Exclusivity among Top-Selling Prescription Drugs in the United States." JAMA Internal Medicine 175 (4): 635–637. https://doi.org/10.1001/jamainternmed.2014.7968.

Within3. 2024. "Essential Guide to the Drug Commercialization Process." Within3, September 10. https://within3.com/guides/essential-guide-to-drug-commercialization-process.

Chapter 4

Allucent. 2024. "BLA vs NDA: Regulatory Differences for Market Approval." Allucent, January 8. https://www.allucent.com/resources/blog/what-are-regulatory-differences-between-nda-and-bla.

Beaney, Abigail. 2024. "FDA Drug Application Costs Set to Rise to $4.3m from October." Clinical Trials Arena, July 31. https://www.clinicaltrialsarena.com/news/fda-cost-revealed-2025-application-drug/.

BioStock. 2023. "Drug Development – The Four Phases." BioStock, January 2. https://www.biostock.se/en/2023/01/drug-development-the-four-phases/.

Brown, Dean G., Daniel G. Wobst, Nicholas M. Hubbard, and Melanie H. Lee. 2022. "Clinical Development Times for Innovative Drugs." Nature Reviews Drug Discovery 21 (11): 793–794. https://doi.org/10.1038/d41573-021-00190-9.

DiMasi, Joseph A., Henry G. Grabowski, and Ronald W. Hansen. 2016. "Innovation in the Pharmaceutical Industry: New Estimates of R&D Costs." Journal of Health Economics 47: 20–33. https://doi.org/10.1016/j.jhealeco.2016.01.012.

DLRC. 2024. "EMA Updated Fees: What Does This Mean for My Applications?" DLRCGroup, August 9. https://www.dlrcgroup.com/ema-updated-fees-what-does-this-mean-for-my-applications/.

Dowden, Helen, and Jamie Munro. 2019. "Trends in Clinical Success Rates and Therapeutic Focus." Nature Reviews Drug Discovery 18 (7): 495–496. https://doi.org/10.1038/d41573-019-00074-z.

European Medicines Agency. 2025. "Fees Payable to the European Medicines Agency." European Medicines Agency, January. https://www.ema.europa.eu/en/about-us/fees-payable-european-medicines-agency.

Federal Register. 2024. "Prescription Drug User Fee Rates for Fiscal Year 2025." Fedcral Register, National Archives and Records Administration, July 31. https://www.federalregister.gov/documents/2024/07/31/2024-16875/prescription-drug-user-fee-rates-for-fiscal-year-2025.

Hinkson, Izumi V., Benjamin Madej, and Eric A. Stahlberg. 2020. "Accelerating Therapeutics for Opportunities in Medicine: A Paradigm Shift in Drug Discovery." Frontiers in Pharmacology 11: 770. https://doi.org/10.3389/fphar.2020.00770.

MarketsandMarkets. 2023. Cell & Gene Therapy Manufacturing Services Market Drivers & Opportunities. MarketsandMarkets, January. https://www.marketsandmarkets.com/Market-Reports/cell-gene-therapy-manufacturing-services-market-180609441.html.

Paul, Steven M., Daniel S. Mytelka, Christopher T. Dunwiddie, Charles C. Persinger, Bernard H. Munos, Stacy R. Lindborg, and Aaron L. Schacht. 2010. "How to Improve R&D Productivity: The Pharmaceutical Industry's Grand Challenge." *Nature Reviews Drug Discovery* 9 (3): 203–214. https://doi.org/10.1038/nrd3078.

PharmaInfo. 2023. "Drug Approval Process in Japan." PharmaInfo, August 29. https://www.m-pharmainfo.com/2023/06/drug-approval-process-in-japan.html.

QLS Advisors. 2021. Clinical Development Success Rates and Contributing Factors 2011–2020. Biotechnology Innovation Organization, February. https://go.bio.org/rs/490-EHZ-999/images/ClinicalDevelopmentSuccessRates2011_2020.pdf.

Roden, Brian. 2023. "The Staggering Cost of Drug Development: A Look at the Numbers." GreenField Chemical Inc., August 10. https://greenfieldchemical.com/2023/08/10/the-staggering-cost-of-drug-development-a-look-at-the-numbers/.

Schlander, Michael, Karla Hernandez-Villafuerte, Chih-Yuan Cheng, Jorge Mestre-Ferrandiz, and Michael Baumann. 2021. "How Much Does It Cost to Research and Develop a New Drug? A Systematic Review and Assessment." *Pharmacoeconomics* 39: 1243–1269. https://doi.org/10.1007/s40273-021-01065-y.

Sertkaya, Aylin, and Calvin Franz. 2022. *Antimicrobial Drugs Market Returns Analysis: Final Report*. Washington, DC: Office of the Assistant Secretary for Planning and Evaluation (ASPE), December 16. https://www.ncbi.nlm.nih.gov/books/NBK603193/.

Sertkaya, Aylin, Trinidad Beleche, Amber Jessup, and Benjamin D. Sommers. 2024. "Costs of Drug Development and Research and Development Intensity in the US, 2000–2018." JAMA Network Open 7 (6): e2415445. https://doi.org/10.1001/jamanetworkopen.2024.15445.

Singh, Debashis. 2004. "Merck Withdraws Arthritis Drug Worldwide." BMJ 329 (7470): 816.2. https://doi.org/10.1136/bmj.329.7470.816-a.

Sun, Duxin, Zhenzhen Tang, Qiang Chen, and Yitao Wang. 2022. "Why 90% of Clinical Drug Development Fails and How to Improve It?" Acta Pharmaceutica Sinica B 12 (7): 3049–3062. https://doi.org/10.1016/j.apsb.2022.02.002.

Takebe, Tohru, Ryoka Imai, and Shunsuke Ono. 2018. "The Current Status of Drug Discovery and Development as Originated in United States Academia: The Influence of Industrial and Academic Collaboration on Drug Discovery and Development." Clinical and Translational Science 11 (6): 597–606. https://doi.org/10.1111/cts.12577.

Waring, Michael J., John Arrowsmith, Andrew R. Leach, Andrew Leeson, Sarah Mandrell, Rolf Owen, Peter Pairaudeau, et al. 2015. "An Analysis of the Attrition of Drug Candidates from Four Major Pharmaceutical Companies." Nature Reviews Drug Discovery 14 (7): 475–486. https://doi.org/10.1038/nrd4609.

Wong, Chi Heem, Kien Wei Siah, and Andrew W. Lo. 2019. "Estimation of Clinical Trial Success Rates and Related Parameters." Biostatistics 20 (2): 273–286. https://doi.org/10.1093/biostatistics/kxx069.

Chapter 5

Baur, Michael, Marco Bühren, Matthias Buente, Morris Hosseini, and Thilo Kaltenbach. 2023. Beyond the Numbers: Key Characteristics of Successful Pharma Companies. Roland Berger, May 24. https://www.rolandberger.com/en/Insights/Publications/Beyond-the-numbers-Key-characteristics-of-successful-pharma-companies.html.

Becker, Zoey. 2023. "Axsome Expands Sales Force to Support Launch of Fast-Growing Depression Med Auvelity." FiercePharma, November 6. https://www.fiercepharma.com/pharma/axsome-inflate-its-sales-team-nearly-100-fast-growing-depression-med-auvelity.

Bleys, Joachim, Werner Rehm, Sean Ryan, and Peter Wright. 2022. "A Biotech Survival Kit for a Challenging Public-Market Environment." McKinsey & Company, September 19. https://www.mckinsey.com/industries/life-sciences/our-insights/a-biotech-survival-kit-for-a-challenging-public-market-environment.

Chaganti, Subba Rao. 2023. Reimagine Pharma Marketing: Make It Future-Proof! PharmaMed Press / BSP Books.

Chang, Steven, Alexandra Young, Mike Davitian, Wyatt Gotbetter, and Dean Giovanniello. 2020. Launch Costs-Spend Wisely: First-in-Class and Follow-On Launch Cost Analysis. Health Advances, March. https://healthadvances.com/application/files/5216/9341/5243/health_advances_launch_costs_spend_wisely.pdf.

Damodaran, Aswath. 2025. "Margins by Sector (US)." NYU, January. https://pages.stern.nyu.edu/~adamodar/New_Home_Page/datafile/margin.html.

David, Frank S., Seth Robey, and Andrew Matthews. 2017. The Pharmagellan Guide to Biotech Forecasting and Valuation. Pharmagellan, LLC.

Davitian, Mike, Haley Fitzpatrick, Grace Perkins, Tara Breton, Remy Denzler, and Dean Giovanniello. 2018. Launch Excellence: Once in a Life Cycle Opportunity. Health Advances, July. https://healthadvances.com/application/files/5516/9341/5245/launch-excellence-life-cycle-healthadvances.pdf.

de Rojas, Carlos. 2024. "Biotech Valuation Multiples: How to Value a Biotech Company?" Labiotech, June 14. https://www.labiotech.eu/expert-advice/biotech-valuation/.

Dunleavy, Kevin. 2025. "Vertex Lays Out Plan to Increase Access and Launch Non-Opioid Pain Reliever Journavx." FiercePharma, February 11. https://www.fiercepharma.com/pharma/vertex-lays-out-plan-increase-access-and-launch-non-opioid-pain-reliever-journavx.

Edwards, Mark. 2017. "Effective Royalty Rates in Biopharma Alliances: What They Are & Why Use Them in Negotiations." les Nouvelles—Journal of the Licensing Executives Society 52 (1). https://ssrn.com/abstract=2904101.

Edwards, Mark. 2019. "Recent Trends in Effective Royalty Rates of Biopharma Alliances." les Nouvelles—Journal of the Licensing Executives Society 54 (4). https://ssrn.com/abstract=3470175.

Frei, Patrik, and Kumlesh K. Dev. 2013. "Drug Dealers: $20 Trillion of In-Licensing Payments." Drug Discovery Today 18 (21–22): 1027–1029. https://doi.org/10.1016/j.drudis.2013.08.011.

Glassdoor. 2025. "Job Search & Career Community." Glassdoor LLC. https://www.glassdoor.com.

Jiang, John, Jing Kong, and Joseph Grogan. 2021. "How Did the Public U.S. Drugmakers' Sales, Expenses and Profits Change over Time?" USC Schaeffer, November 5. https://healthpolicy.usc.edu/evidence-base/how-did-the-public-u-s-drugmakers-sales-expenses-and-profits-change-over-time/.

Kornfield, Rachel, Julie Donohue, Ernst Berndt, and G. Caleb Alexander. 2013. "Promotion of Prescription Drugs to Consumers and Providers, 2001–2010." PLOS One 8: e55504. https://doi.org/10.1371/journal.pone.0055504.

Lee, Namryoung. 2021. "The Association between the Selling, General & Administrative Expenses and Age at IPO of Biotech Companies." Academy of Accounting and Financial Studies Journal, Allied Business Academies. https://www.abacademies.org/articles/the-association-between-the-selling-general--administrative-expenses-and-age-at-ipo-of-biotech-companies-10273.html.

Locust Walk Institute. 2017. "Biopharma Valuation Analysis." Locust Walk Institute, September. https://www.locustwalk.com/wp-content/uploads/2017/09/Biopharma-Valuation-Analysis.pdf.

Neuendorf, Elias, Kara O'Connell, Patrik Frei, and Kumlesh Dev. 2023. "A Formula for Drug Licensing Deals." Nature News, June 1. https://www.nature.com/articles/d43747-023-00062-8.

O'Connell, Kara E., Patrik Frei, and Kumlesh K. Dev. 2014. "The Premium of a Big Pharma License Deal." Nature Biotechnology 32 (7): 617–619. https://doi.org/10.1038/nbt.2946.

Rottgen, Raphael. 2025. "Biotech Valuation Idiosyncrasies and Best Practices." *Toptal Management Consultants*. https://www.toptal.com/management-consultants/valuation/biotech-valuation.

Stasior, Jonathan, Brian Machinist, and Michael Esposito. 2018. *Valuing Pharmaceutical Assets: When to Use NPV vs rNPV*. Alacrita Pharma & Biotech Consulting, August. https://www.alacrita.com/whitepapers/valuing-pharmaceutical-assets-when-to-use-npv-vs-rnpv.

Thompson, Charlie. 2024. "Pharma Sales Force Sizing Strategy – Classic Approaches." Axtria. https://insights.axtria.com/blog/pharma-sales-force-sizing-strategy-part-1-classic-approaches.

Wall Street Prep. 2024a. "Working Capital." *Wall Street Prep*, July 12. https://www.wallstreetprep.com/knowledge/working-capital/.

Wall Street Prep. 2024b. "Capital Expenditure (CAPEX)." *Wall Street Prep*, April 21. https://www.wallstreetprep.com/knowledge/capital-expenditure-capex/.

Wall Street Prep. 2024c. "Depreciation and Amortization (D&A)." *Wall Street Prep*, June 19. https://www.wallstreetprep.com/knowledge/depreciation-vs-amortization/.

Chapter 6

American Cancer Society. 2024. "What Is Lung Cancer?" *American Cancer Society*, January 29. https://www.cancer.org/cancer/types/lung-cancer/about/what-is.html.

Ghoshal, Malini. 2025. "Dosage Details for Keytruda (Pembrolizumab)." *Healthline*, April 23. https://www.healthline.com/health/drugs/keytruda-dosage#how-its-given.

Kantarjian, Hagop, and Yogin Patel. 2017. "High Cancer Drug Prices 4 Years Later—Progress and Prospects." *Nature Reviews Clinical Oncology* 123 (8): 1292–1297. https://doi.org/10.1002/cncr.30545.

Keytruda 2025a. "Keytruda: Non-Small Cell Lung Cancer." *Merck & Co.* https://www.keytruda.com/non-small-cell-lung-cancer/.

Keytruda 2025b. "Keytruda: Cost, Insurance & Financial Help." *Merck & Co.* https://www.keytruda.com/financial-support/.

Kraus, Sebastian, and Mathias Schmid. 2024. "What Is the Optimal Treatment Duration of Immunotherapy for NSCLC?" *healthbook TIMES Oncology Hematology* 19 (1): 50–55. https://doi.org/10.36000/hbt.oh.2024.19.139.

Reinmuth, Niels, Nadine Payer, Thomas Muley, et al. 2013. "Treatment and Outcome of Patients with Metastatic NSCLC: A Retrospective Institution Analysis of 493 Patients." *Respiratory Research* 14: 1–9. https://doi.org/10.1186/1465-9921-14-139.

Schouten, Arianna. 2025. *Addressing the Financial Implications of PD-1/PD-L1 Immune Checkpoint Inhibitors: A Policy Analysis of Access and Inclusion on the WHO Model List of Essential Medicines*. World Health Organization, January. https://cdn.who.int/media/docs/default-source/2025-eml-expert-committee/addition-of-new-medicines/a.22-pd1-pdl1-icis_financial-impact-report.pdf?sfvrsn=7378e942_1.

SEER. 2025. "Cancer Stat Facts: Lung and Bronchus Cancer." *National Cancer Institute, Surveillance, Epidemiology, and End Results Program (SEER).* https://seer.cancer.gov/statfacts/html/lungb.html.

Wilder, Kristie. 2024. "New 2024 Population Estimates Show Nation's Population Grew by About 1% to 340.1 Million since 2023." *United States Census Bureau,* December 18. https://www.census.gov/library/stories/2024/12/population-estimates.html.

Appendices

Aggarwal, Ankul and Nitin Bahl. 2021. *Start-Ups and Early Stage Companies: A Valuation Insight.* KPMG. https://assets.kpmg.com/content/dam/kpmg/kw/pdf/insights/2021/05/valuation-startup-web.pdf.

Beaton, Neil, and John Sawyer. 2019. "Use of Monte Carlo: Simulations in Valuation." AIRA *Journal* 32 (2). https://www.alvarezandmarsal.com/sites/default/files/88413_val_monte_carlo_aira_article.pdf.

Biotechgate. 2025. "Biotechgate." *Venture Valuation.* https://www.venturevaluation.com/about-biotechgate/.

Damodaran, Aswath. 2025a. "Revenue Multiples by Sector (US)." January. https://pages.stern.nyu.edu/~adamodar/New_Home_Page/datafile/psdata.html.

Damodaran, Aswath. 2025b. "Enterprise Value Multiples by Sector (US)." January. https://pages.stern.nyu.edu/~adamodar/New_Home_Page/datafile/vebitda.html.

Department for Science, Innovation & Technology. 2023. *Independent Review of University Spin-out Companies.* Crown, November. https://assets.publishing.service.gov.uk/media/6549fcb23ff5770013a88131/independent_review_of_university_spin-out_companies.pdf.

Edwards, Mark. 2019. "Milestone Payments in Biopharma: Negotiating an Equitable Value Allocation." *Nature News*, May 29. https://www.nature.com/articles/d43747-020-00675-3.

Fein, Adam. 2023a. "Four Trends That Will Pop the $250 Billion Gross-to-Net Bubble—and Transform PBMs, Market Access, and Benefit Design." *Drug Channels*, April 4. https://www.drugchannels.net/2023/04/four-trends-that-will-pop-250-billion.html.

Fein, Adam. 2023b. "Exclusive: The 340B Program Reached $54 Billion in 2022—Up 22% vs. 2021." *Drug Channels*, September 24. https://www.drugchannels.net/2023/09/exclusive-340b-program-reached-54.html.

Guth, Alex, Nikita Jain, and Vivek Venkataramani. 2024. "Refining Gross-to-Net Expectations for Improved Strategic Planning." *L.E.K. Consulting*, November 4. https://www.lek.com/insights/hea/us/ei/refining-gross-net-expectations-improved-strategic-planning.

Intellectual Property Research Associates. 2022. *Royalty Rates for Pharmaceuticals & Biotechnology, 9th Edition.* Intellectual Property Research Associates. https://techpipeline.com/product/royalty-rates-for-pharmaceuticals-biotechnology-9th-edition/.

Neuendorf, Elias, Christoph Grundel, and Nicolas Furet. 2023. "A Formula for Drug Licensing Deals." *Nature News*, June 1. https://www.nature.com/articles/d43747-023-00062-8.

Prasad, Kenath. 2025. "Startups, Science & IP: Intellectual Property & Licensing 101 for Deep-Tech Founders." *Medium*, May 5. https://medium.com/included-vc/startups-science-ip-intellectual-property-licensing-101-for-deep-tech-founders-0685a13851fd.

Rottgen, Raphael. 2025. "Biotech Valuation Idiosyncrasies and Best Practices." *Toptal Management Consultants.* https://www.toptal.com/management-consultants/valuation/biotech-valuation.

Seeley, Elizabeth. 2022. "The Impact of Pharmaceutical Wholesalers on U.S. Drug Spending." *The Commonwealth Fund*, July 20. https://www. commonwealthfund.org/publications/issue-briefs/2022/jul/impact-pharmaceutical-wholesalers-drug-spending.

Sood, Neeraj, Tiffany Shih, Karen Van Nuys, and Dana Goldman. 2017. *The Flow of Money through the Pharmaceutical Distribution System*. USC Schaeffer, June. https://schaeffer.usc.edu/wp-content/uploads/2024/10/The-Flow-of-Money-Through-the-Pharmaceutical-Distribution-System_Final_Spreadsheet.pdf.

Stanford Office of Technology Licensing. 2025. "OTL's Process." *Stanford University*. https://otl.stanford.edu/researchers/otls-process.

Resources

Sertkaya, Aylin, Trinidad Beleche, Amber Jessup, and Benjamin D. Sommers. 2024. "Costs of Drug Development and Research and Development Intensity in the US, 2000–2018." *JAMA Network Open* 7 (6): e2415445. https://doi.org/10.1001/jamanetworkopen.2024.15445.

About the Authors

Tristan Yeung, MD, is a Management Consultant at AGMI Group specializing in biotech asset valuation and due diligence. He earned his MD from Harvard Medical School, where he engaged extensively with the Harvard Innovation Labs, HMS Makerspace, Massachusetts Eye and Ear's Retina Imaging Lab, and Massachusetts General Hospital's COVID-19 Radiology Research Group. He completed his internal medicine internship at California Pacific Medical Center, a Sutter Health institution academically affiliated with UCSF. Dr. Yeung's past research includes translational work at Stanford's Department of Radiology and Canary Center, where he implemented minimally invasive imaging and novel microRNA-encapsulated nanoparticle drug delivery systems for localized cancer treatment. He holds a B.S. in Bioengineering with Distinction from Stanford University, where he was a Terman Engineering Scholar and developed a vitamin-eluting subdermal implantable rod to treat nutritional deficiencies in post-bariatric surgery patients. Dr. Yeung leverages his distinctive blend of clinical training, translational research expertise, and entrepreneurial vision to advance biomedical innovation and drive solutions to unmet needs in business and healthcare.

Jean Cruz, PhD, is the Co-founder and Global Managing Partner of AGMI Group, advising pharmaceutical companies, investors, and biotech founders on R&D strategy, M&A, and global market expansion. He is also a scientist, strategist, and investor with deep expertise in biopharma innovation and private capital. Dr. Cruz earned his PhD in Biomedical Engineering from Cornell University, where he led multi-institutional research initiatives in neurodegeneration and therapeutic imaging. In 2017, he joined Harvard Medical School and Massachusetts General Hospital as a postdoctoral fellow and junior faculty member, contributing to first-in-human diagnostic technologies and publishing in journals such as Nature. Dr. Cruz later served as an Engagement Manager at McKinsey & Company, where he led high-impact valuation and portfolio strategy efforts for neuro, immunology, and cell and gene therapy assets across the U.S. and Latin America.

Yen-Po (Harvey) Chin, MD, PhD, is the Co-founder and Chairman of AGMI Group and a recognized authority in life sciences business strategy and asset valuation. With a unique background spanning clinical medicine, advanced analytics, and top-tier management consulting, he brings a rare blend of scientific depth and commercial insight to the healthcare and biotech industries. As Chairman of AGMI Group and a former management consultant at McKinsey & Company, Dr. Chin has led critical projects in licensing and M&A, cross-border fundraising, and post-deal integration for global pharmaceutical companies and private equity investors. He has operated across the full investment lifecycle—from due diligence and risk-adjusted valuation modeling to term sheet design—earning a reputation for analytical rigor and strategic impact at the boardroom level. Dr. Chin is an inaugural HealthTech Innovation Fellow at Harvard Medical School and was named to the Forbes 30 Under 30 list. He holds a postdoctoral master's degree in biomedical informatics from Harvard Medical School, a PhD in biomedical informatics from Taipei Medical University, and an MD from National Yang Ming Chiao Tung University in Taiwan. Passionate about mentoring the next generation of leaders at the intersection of science, finance, and strategy, Dr. Chin has guided over 1,000 students and professionals worldwide through his YouTube channel and online coaching programs.

www.ingramcontent.com/pod-product-compliance
Lightning Source LLC
Chambersburg PA
CBHW061128220326
41599CB00024B/4206